T0134056

ENERGY MIRACLES

ENERGY MIRACLES

The Global Warming Backup Plan

H. B. Glushakow

Published by

Jenny Stanford Publishing Pte. Ltd.
Level 34, Centennial Tower
3 Temasek Avenue
Singapore 039190

Email: editorial@jennystanford.com
Web: www.jennystanford.com

British Library Cataloguing-in-Publication Data
A catalogue record for this book is available from the British Library.

Energy Miracles: The Global Warming Backup Plan

ISBN 978-981-4968-18-8 (Hardcover)
ISBN 978-1-003-28442-0 (eBook)

This book is dedicated to my wife, Sharon, without whose love and gargantuan support I would never have been able to complete this project. The few times I ground to a halt, it was she and her smart suggestions that put me back on the road toward Energy Miracles. Also, to Martin Luther King, Jr., who taught me that one need not be a perfect man in order to dream great dreams and turn those dreams into actuality. And to Ron Hubbard whose interest in energy and molecular phenomena sparked my own.

Contents

Foreword

Energy Miracles: The Global Warming Backup Plan is a refreshing, non-political, novel approach to identifying both climate and energy issues of the world and offering a way to reverse and remedy them.

This book brutally and efficiently cuts through all the non-scientific hoopla often shoved in our faces on a daily basis, and presents us with the facts regarding our current, bleak state of polluted waters, air, and atmosphere. It shows the actual effects (such as rising seas) caused by emissions from power plants burning fossil fuels that we've, through the ages, come to depend on.

I'm no Einstein, but luckily, one doesn't have to be one to appreciate this book. The author nicely pooled together a cross section of the scientific/electrical history that got us here in the first place, including the brilliant history of the world's greatest scientists, researchers, and inventors as well as the paths of some of the misguided ones. And then he presents a way to innovate and lead the world to scientific victory over the problems caused by the current ways of producing electricity.

I've been an electrical engineer for 30 years and I can say the fantastic conveyance of electrical facts and history contained in this book accurately identifies why things went wrong in the electrical-discovery world. The book is right!!!! We didn't learn diddly-squat about making electricity in college. We got taught only what to do with it once we had it. The actual operation of **making** electricity has always been taken for granted. We have a receptacle. We plug something in, and it works. I think most people trying to solve climate change today don't think beyond the receptacle.

Mr. Glushakow has left me inspired by not only highlighting energy's history but shining a spotlight on the path to discoveries yet to be. Keep your eyes open as energy miracles are inevitable and may be helped along by Mother Earth herself. Often, more so than not, the answers are right in front of us.

The book should open a lot of eyes and though it is written to be understandable by general audiences, it should be made mandatory reading for all physics students and electrical engineers.

Dion Neri
Senior Electrical Engineer
MCG Electronics Inc.
New York, USA

Introduction

I wrote this book because despite 10 years of massive effort and trillions of dollars of investment into wind and solar energy, carbon emissions continue to increase, the planet continues to heat up, and the seas continue to rise. Global warming has caught the attention of most of the people of the Earth. Many are afraid and looking for answers, but at the same time, they are unconvinced by solutions that tell them all will be well if they simply recycle their trash, take shorter showers, stop eating beef, drive electric vehicles, manufacture more solar panels, or wait for BP to come to the rescue.

Wind and solar energy are the primary technologies supported by the Paris Climate Accords, and as they are clean sources of energy that can produce new jobs at the same time as reducing carbon emissions, they are important to support. The decade from 2010 to 2020 was a busy one for both these technologies, with each receiving over a trillion dollars in investment funds, which quadrupled their global power output. Good as that record sounds, their total output still accounts for only 3% of the globe's yearly power consumption. Bill Gates may have been the first to warn that a more comprehensive solution was needed when he stood up at the TED conference in 2010 and coined the term Energy Miracle to signify a new 21st-century source of energy that was capable of meeting the energy needs of the planet without the destructive effects that result from burning fossil fuels.[1] It was his belief then, and remains so today, that innovation is needed to support the wind and solar industries because, alone, they are not capable of providing the volume of clean energy that the planet so desperately needs. In the past decade, he has taken a lot of heat from naysayers who for various reasons did not appreciate his remarks, but as will be seen in this book, the foresight of Mr. Gates has been vindicated. We may have lost a decade, but the need for immediate innovation remains an urgent priority.

---

[1]TEDs (Technology, Entertainment, Design) are yearly conferences, attended by industry leaders and impactful people, which have been described as journeys into the future in the company of those creating it.

When I first heard Bill Gates saying the world needs an Energy Miracle, I realized that I had already unraveled a major part of the mystery that had so far prevented such Energy Miracles from materializing.

I am a Senior Member of the IEEE and for several years, I have been researching the major discoveries in electricity. In the course of that study, I discovered several amazing facts: One of them is that the last major electrical discovery occurred in 1938. It was as though someone turned off the electrical discovery switch at that precise instant, and ever since, we have been stuck with the same 19th-century power sources we were using 100 years ago. I wanted to know why. In looking closer I identified five key scientific fundamentals that had informed and underpinned almost every one of history's top 42 electrical discoveries. This led to the second amazing thing: All five of those fundamentals were abandoned in the mid-1930s. Every one of them. These included the very definition of energy itself, what gives energy its impetus, the structure required for energy to occur, the requirement of a medium for energy to propagate through, and the mechanism by which energy propagates through that medium. All five were abandoned in one fell swoop, replaced by theories of quantum mechanics. Since then, not a single new advance has been made in electrical energy production. This book is the first to sound the alarm on this situation.

Yes, I said quantum mechanics. But wait. Every book on global warming is written with some degree of complexity and technology. I was forced to address some aspects of quantum mechanics after discovering how much it was directly impeding the discovery of Energy Miracles, but it is done in clear terms, without resort to mathematics, and I believe I have made its darkest weeds comprehensible to the general public for the very first time.

As you will come to see in the first part of this book, we cannot effectively take on the effects of global warming without a 21st-century Energy Miracle. The quest for this new source of energy will serve as a backup to all the actions currently being undertaken to deal with climate change. A two-pronged approach is proposed: one, scientific; the other, political.

Bill Gates took a similar approach in his thoughtful book, *How to Avoid a Climate Disaster: The Solutions We Have and the Breakthroughs We Need,* released in mid-2021; only he did not know and was unable to predict where such an Energy Miracle could be found. It is to that exact pursuit that the book you are now reading is addressed. If your plan is to discover an Energy Miracle, you ought to focus on energy. Armed with the knowledge contained in these pages, one can genuinely wax optimistic about our chances of slaying the climate-change dragon.

In this book you will learn for the first time not just how we became stuck with 19th-century energy technology, but what can be done about it. The book introduces, in some detail, those five aforementioned keys to the Energy Miracles and proposes their use in 12 of the most fertile areas of energy research, including thermoelectrics, superconductivity, Nikola Tesla's Free Energy, nuclear fission and fusion, and more.

As for the political side, the Paris Climate Accords is an agreement drafted in 2015 and subsequently signed by close to 200 countries. For 200 countries to come together and agree on *anything* is staggering; the momentum should not be wasted. All global warming mitigation projects in existence or in planning should be strongly supported and even strengthened. But none of those initiatives are designed to directly create an Energy Miracle, and so Chapter 9 of this book introduces **The Energy Miracle Challenge**, an international game to generate new spark into the Paris Climate Accords. This is a game that every country can play, any corporation can play, and even motivated individuals can play. It provides incredible incentives for those who enter, but the biggest incentive of all is that the resulting Energy Miracles will be made available free of charge to all the peoples of Earth.

Dealing with global overheating is a fiery urgency. The mitigation of its imminent consequences is the primary goal of this book. The warming of the planet would not be such a pressing problem except for the fact that humans in their wisdom have chosen to locate one out of every eight people in low-lying coastal areas. More than half of the largest cities on the planet are situated on a coast. What happens to those one billion people when the seas begin to rise? It does not matter a hoot whether or not global warming is the result of human-caused action. What matters is that **solutions** to global warming will **only** be the result of human-caused action.

New sources of energy (Energy Miracles) have another benefit. Aside from slowing or reversing global warming and its planet-wide effects, these same new sources of energy will solve many of the other problems plaguing the planet, including air pollution, hunger, and lack of fresh water.

Energy innovation requires scientists, and for science to successfully meet the challenges of global warming requires the return to a more logical approach to solving problems. It will rely on hands-on experiments, like the ones Benjamin Franklin did with his kite. Such experimentation assisted the scientists of old to see things for what they were and then investigate how those things worked. Take climate. Before Franklin, the world was flummoxed by why ships took weeks longer to make the trip from Europe to America than on their return journey. Franklin noticed it on his frequent transatlantic trips. Through experimentation and careful accumulation of facts, he was able to discover and map the Gulf Stream. This discovery led to the ability to understand the causes of currents and weather patterns and allowed European captains to map better routes and take on enough food to provide for their crews on the return trips to Southampton or Marseilles.

Science, to be of any use, must be understandable and capable of taming the confusions of the universe for the benefit of humankind. With respect, modern science has largely abandoned that road. The norm has become the concoction of theories so obtuse that scientists brag there is no one on the planet capable of understanding them (including the scientists themselves). Currently we have science telling us that all that is real is in the quantum world, which we can neither see nor understand, and the most likely way to discover anything of value is to get someone to build you a five-billion-dollar particle accelerator. Those things are not true, and the proof is they have got us not a whit closer to solving civilization's primary challenge: global warming. To be sure, quantum mechanics will continue to have its place. But in the scheme of things, and in keeping with its name, it will be a very small place, indeed.

Bill Gates is not wrong to be optimistic about our chances of surviving a climate disaster. Something can be done, but to proceed effectively, one must proceed with an open mind. In the past, nearly every pioneer of knowledge has at some time been judged a quack or worse. Galileo was put under house arrest for the last 9 years of his life for suggesting the Earth moved around the sun. The original discoverers of steam energy, electricity, artillery, and astronomy were all considered of little or no consequence and belittled by their peers; Martin Luther King was pronounced a subversive during his life, and

Ron Hubbard, irrelevant, while Nikola Tesla was disregarded as a wacky madman. In these days of Covid-19 extreme hand-washing, it is instructive to reflect on the work of Dr. Ignaz Semmelweis, the Hungarian physician and scientist who in the 1840s discovered that a doctor or midwife could reduce the number of women who died in childbirth from 1 in 10 to less than 1 in 100 by the simple action of the doctor washing his hands. Semmelweis published his results in a book in 1858, but they were rejected, and he was widely attacked by the European medical community for daring to insinuate that their dirty hands were causing so much death. For many decades thereafter, tens of thousands of women died needlessly until his advice was finally duplicated and acted upon. As regards to global warming, the Earth may not have the luxury of those decades.

In 1997, John Horgan, senior science editor of the *Scientific American*, wrote a book entitled *The End of Science* in which, dismayed after interviewing too many leading scientists, he wrote, "If one believes in science, one must accept the possibility—even the probability—that the great era of scientific discovery is over." The book you are now reading will show that this fatalism is far from the truth. A new age of scientific discovery is just beginning. At least it better be.

Discovering an Energy Miracle will change everything.

H. B. Glushakow
Senior Member IEEE

A note to the scientists: Not to say that big problems can be solved with a finger snap, but advancement is always possible. What does kind of slow you down, however, is when you are forced to start with a black hood over your head, both arms tied behind you, and legs shackled with chains. This book can remove the black hood and free your arms, but the rest is up to you.

What Is an Energy Miracle?

"If a genie offered me one wish, a single breakthrough in just one activity that drives climate change, I'd pick making electricity."[1]

— Bill Gates

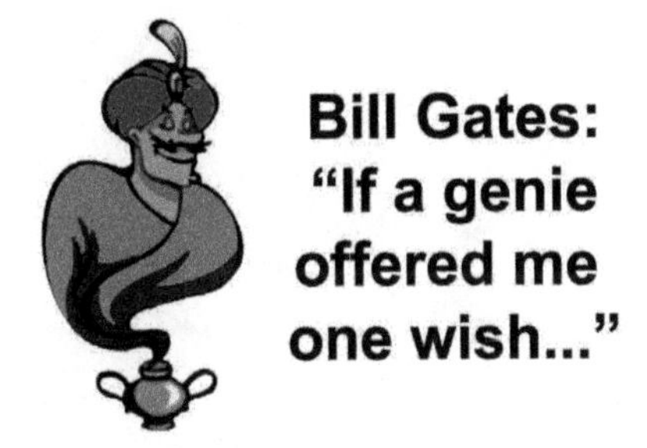

Figure 1 Bill Gates' most fervent wish.

The climate is changing, and Bill Gates coined the term Energy Miracle and exactly defined its conditions in the hope that the world would innovate ahead of the negative effects of that change. He was the first to bring attention to the reality that in order to get to grips with the consequences of global warming, new inventions would be needed, with new sources of electricity at the top of his wish list. When Mr. Gates talks about the need for an Energy Miracle, he is not referring to something that is impossible. He points out that in the course of his career, he has seen many such miracles emerge as the result of research and development and the native human capacity to innovate. Miracles like the personal computer, the Internet, and polio vaccine did not just happen by accident—they were the result of dedicated purpose and hard work.

A Gates 21st-century Energy Miracle must meet three conditions:

1. It must be cheaper than today's hydrocarbon energy,
2. It must have zero CO_2 emissions, and
3. It must be as reliable as today's overall energy systems.

[1]Bill Gates, How to Avoid a Climate Disaster: The Solutions We Have and the Breakthroughs We Need, Alfred A. Knopf, 2021. Quote used with permission.

And when you put all those requirements together, you can only say: "WOW!

We need an energy miracle." [2]

The stakes are high, and time is not on our side. Every day the consequences of climate change grow more ominous.

---

[2]Bill Gates in Gates Notes. Copyright © The Gates Notes LLC and used under Creative Commons license.

Part I

Preface to Part I

In the first part of this book, we introduce the roots of the climate change idea and the documented but often unappreciated real-world effects of burning fossil fuels (coal, oil, and gas). We also take an objective look at the results of trying to replace those fuels—the progress (or lack of it) that has been made so far.

This morphs into the subject of electrical energy. If you want to find out what is plaguing attempts at finding new sources of energy, a good place to look is the subject of energy itself. This includes a survey of history's all-time greatest discoveries in electrical energy and what exactly happened to derail the subject of electrical engineering.

Finally, the five keys to the Energy Miracles are set forth. These are the basic fundamentals that informed and underpinned just about every one of the major discoveries in electrical energy but were precipitously abandoned in the early part of the 20th century. With those five keys rehabilitated, a path to new Energy Miracles is assured.

With the path thus set, the question is then asked, how can the world get its best minds and limited resources all moving ahead in the same direction to solve the climate change problem? The first part of this book ends with a mechanism proposed to exactly answer that question at the same time bringing new spark and hope back into the Paris Climate Accords.

Chapter 1

Climate Change

Climate: What Is It?

It is no wonder that the subject of climate change has become such a pesky one when even the two-volume *Encyclopedia of Weather and Climate* will not venture a clear definition of the word but instead suggests it can only be properly understood in terms of "the arrangement of climates according to their most important characteristics in order to provide each type with a short, unambiguous name or title by which it can be known." Sort of like trying to define "monkey" as "any of the many different types of monkey."

Climate does have a meaning. It comes from the Greek *klimat*, meaning *to lean in a certain direction.* Its first scientific use was in the field of geography where it designated a specific part of the Earth's surface bounded by two circles parallel to the equator where the longest day on one of the circles was exactly 1/2 hour shorter than the longest day on the next circle (see Fig. 2). That concept enabled scientists to discuss different conditions existing within a particular "climate": how cold, how humid, how much wind, quality of air, wind direction, etc.

But that was a long time ago. By the end of the 18th century, scientists had dispensed with the "length of day" requirement, and the term had morphed into a more popular usage, referring

Energy Miracles: The Global Warming Backup Plan
H. B. Glushakow

ISBN 978-981-4968-18-8 (Hardcover), 978-1-003-28442-0 (eBook)
www.jennystanford.com

to *any tract of land, region, or country, differing from another with regard to temperature of the air, the seasons, or peculiar qualities.*[4] Thus we could talk about a warm or cold climate, a dry climate, a mountainside climate, or a noxious climate. The term further evolved to identify "*the prevailing environmental conditions of a place*" before meteorologists began defining it as the "generally prevailing weather conditions of a region."[5]

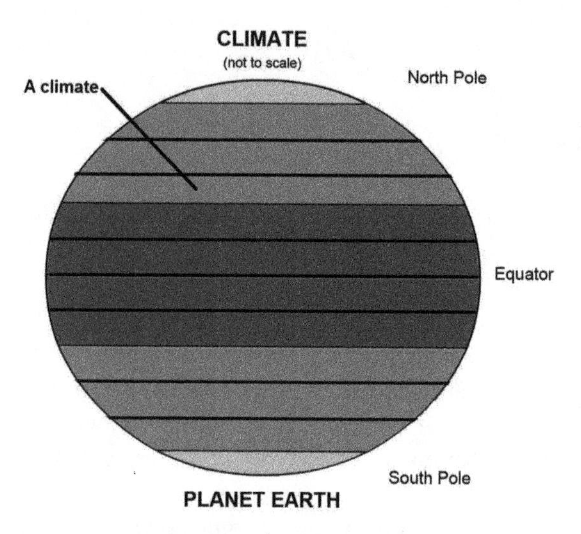

Figure 2 Climate-original definition.

Climate is a big subject, and to deal with it you need to think in larger terms. "*The prevailing environmental conditions of a place*" remains its best definition, for it allows us to incorporate the many critical factors intrinsic to climate that need to be addressed and solved. Much as you need to know something about lions to tame them, so must you know something about climate to tame it.

Climate change does not only mean more super storms and melting glaciers. It also encompasses changes in water and air. Less water and rising seas alike constitute climate change as does changing air quality, whether that change be more arid, frigid, thin, or an atmosphere bursting with poisonous particles. In the 1700s, atmospheric pressure, relative humidity, and air quality indexes

[4]*Noah Webster Dictionary*, 1827.

[5]*Webster's Encyclopedic Unabridged Dictionary of the English Language*.

were not included in definitions of climate simply because they could not be accurately measured. Today, those plus other things like water quality and air quality must be included in discussions of climate since they can now be so accurately measured and have such major impact upon our lives.

Figure 3 Climate tamer.

The aspect of climate most evident in discussions of climate change is temperature. That the world is heating up at an unusually rapid pace is no longer in question. This warming trend is the aspect of climate that we call global warming. When the temperature rises high enough to melt the glaciers, the seas will rise. It is a catastrophe in the making.

Figure 4 is a graph from NASA showing a steady 100-year temperature rise.

Statistics aside, photos and videos of melting glaciers and classically white-crowned mountains that have lost their snowy covers should have put an end to any serious debate on that score. Owners of Alpine ski resorts will tell you they had to produce five times as much artificial snow in 2021 to keep their slopes open for the season than was necessary in 2000 and the warmer weather has already forced ski resorts at lower elevations to close entirely. It is becoming obvious to even the oblivious that the increasing number of super storms, wild fires, severe floods, and storm surges are likely to be related to this warming trend. It is not surprising that this trend has caught the attention of the world when rising seas can threaten an entire country such as Vietnam with extinction. The people in Calcutta (Kolkata) and Mumbai are no less anxious to wake up one day to find themselves under water than the people in Miami and New York.

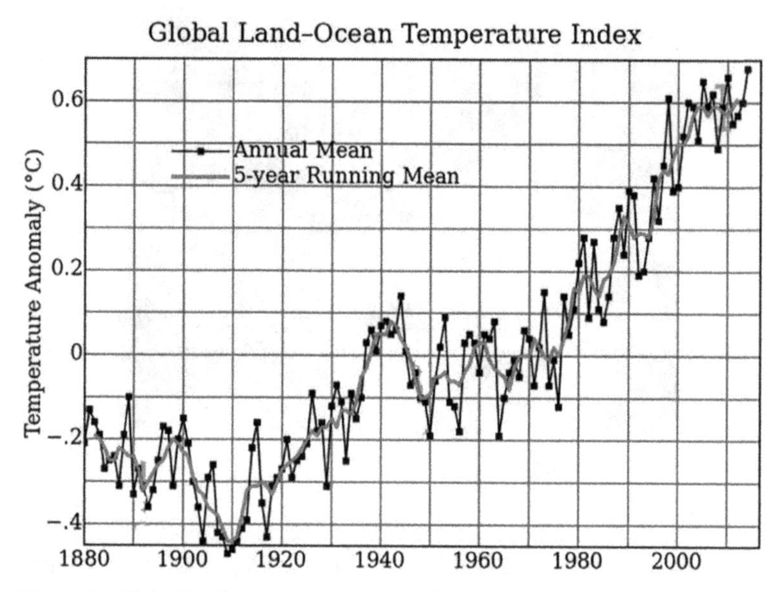

Figure 4 Global land-ocean temperatures by year.

Al Gore: Climate Guru

The world of 2021 knows Al Gore as the planet's Climate Guru. A former Vice President (and almost President) of the U.S., he is the person who has done the most to alert the planet's current generation to the dangers of climate change. For this he was awarded a Nobel Peace Prize. According to Gore, climate change is the world's most dangerous threat.

I went to see Gore's *Inconvenient Truth* movie in 2006. Although touched by the visuals of melting glaciers, etc., what really grabbed me were the NOAA graphs shown in the movie. The first two in Fig. 6 show 1000 years of rising CO_2 concentrations compared to the same 1000 years of rising temperatures on Earth. The fact that the temperature changes so exactly follow changes in CO_2 concentration is too much of a coincidence for a link to be dismissed out of hand.

Figure 5 Al Gore, Climate Guru.

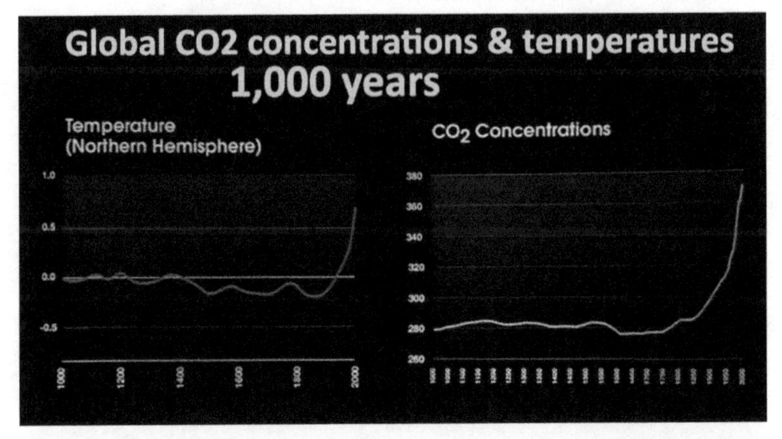

Figure 6 Global temperature and CO_2 concentrations for the last 1000 years.

But 1000 years is pretty short as things go on this planet. What about hundreds of thousands of years? The NOAA graphic in Fig. 7 shows that the average of the eight highest peaks of CO_2 concentration over the last 800,000 years was 266 ppm; and only once (375,000 years ago) has the CO_2 ppm (parts per million) concentration hit 300.

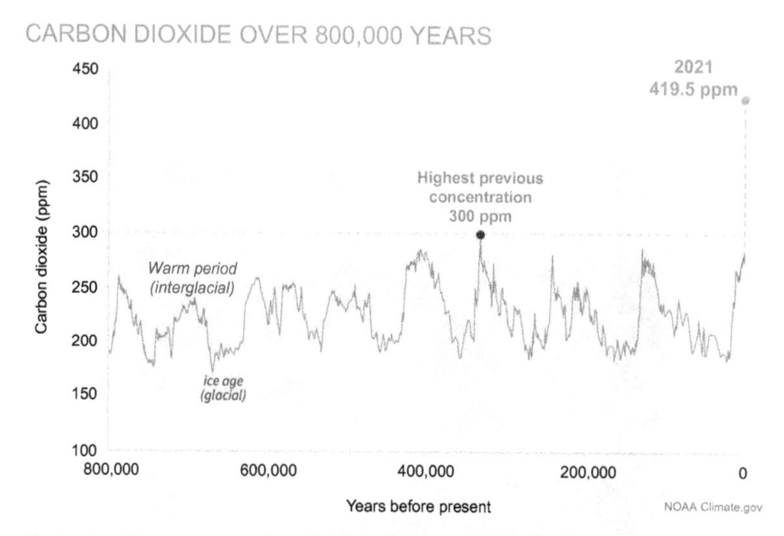

Figure 7 CO_2 concentrations for the last 800,000 years.

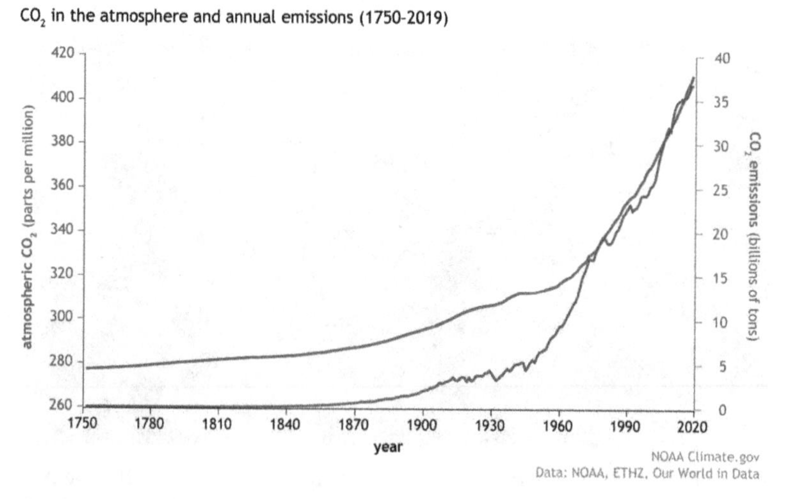

Figure 8 Fossil fuel emissions versus carbon levels in atmosphere.

In 2006, it was 385 with Vice President Gore predicting it would be off the graph within 50 years if fossil fuel emissions were not curtailed. Today, in May 2021, it is 419.5, up an additional 9% in only 15 years and 50% above the pre-industrial levels despite the dip in emissions during the Covid-19 pandemic. This is the highest concentration of carbon dioxide the planet has seen in at least 3 million years.

This final graph (Fig. 8) shows what is driving the high CO_2 concentrations in the atmosphere. **It is exactly coincident with the amount of carbon emissions from fossil fuel generators.**

Short History of Climate Change

As brilliant as he was in alerting the planet to the dangers of global warming, Vice President Gore did not originate the idea. The concept of global warming and the debate about what role, if any, humans may play in it have been going on since the 19th century when scientists were already theorizing that glacial periods and other great climatic changes could be explained by the ability of carbon dioxide in the atmosphere to absorb infrared radiation bouncing off the Earth's surface. The Swedish chemist Svante Arhenius in 1897 calculated that doubling the percentage of carbon dioxide in the air would increase the Earth's temperature by 4°C. But it was not until the release of the papers of British Engineer Guy Stewart Callender (published between 1938 and 1961) that it became clear that humankind was very likely interfering in the otherwise slow-moving carbon dioxide cycle. He calculated that we were "throwing some 9,000 tons of carbon dioxide into the air each minute" as early as 1938, and that in the previous 50 years fuel combustion of fossil fuels had generated 150 million tons of carbon dioxide, three-quarters of which had remained in the atmosphere. His conclusion, from his 1939 paper, "From the best laboratory observations it appears that the principle result of increasing atmospheric carbon dioxide...would be a gradual increase in the mean temperature of the colder regions of the Earth."[6]

By the late 1940s, "Global Warming" was already a public concern. Worries about rising sea levels, forced changes in agriculture, loss of habitats, the melting of Greenland's ice cap and other glaciers, and the potential migration of millions of people being displaced by

---

[6]G. S. Callender, The composition of the atmosphere through the ages, *Meteorological Magazine*, **74**, 1939, 38.

climate change were being voiced in scientific and popular journals. In 1950, *The Saturday Evening Post* ran an article entitled "Is the World Getting Warmer?"

Although some, then as now, discounted the notion that there is anything to fear from the climate, and that climate change was a hoax, a look at either of Gore's movies could have been enough to alert anyone with eyes, ears, and the ability to feel a wind gust in his or her face that something is very not right with our climate. There is change going on that bodes nothing but ill for the peoples of Earth. Two of the most imperiled big cities immediately threatened are both in Florida: Miami and Tampa. Gore's second movie includes video footage of flooded downtown Miami streets with fish swimming alongside submerged cars. That film, *An Inconvenient Sequel: Talking Truth to Power*, was nothing less than prophetic. Released in theaters on July 28, 2017, Gore predicted that storms would now be coming in larger flavors and there would be more of them. Less than a month later super storm Harvey ravaged Texas.

Less than a week after that, Inconvenient Irma, a super storm that dwarfed even Harvey, took center stage plowing its way through the state of Florida and causing billions of dollars of property damage. Scarcely less than one week after that, Hurricane Maria hit Puerto Rico blowing the tops off 90% of the houses and destroying 100% of the electrical grid, which threw the entire country into darkness. All three of those hurricanes had sustained winds of over 111 mph. There have never been three such storms in a single season to hit the U.S. since records have been being kept. Surprisingly, only a few deaths were directly attributed to those back-to-back super storms in the summer of 2017. Forewarned by the latest weather technology, potential casualties were able to either make preparations or flee in advance of those hurricanes. The biggest intervention in peoples' lives turned out to be power outages. The first of those hurricanes (Harvey) interrupted electrical power for close to a million people in Texas, while the second (Irma) knocked out electrical power for 16 million people in the southern U.S., including Florida.

For the first time in 300 years, there were no inhabitants on the island of Barbuda when every human was forced to flee. Houses could be rebuilt, but not without power. Without power, the island had no water, no lights, no Internet, no phone service, no credit cards, no medical treatment. Puerto Rico, which lost all of its power for months, when hit by Hurricane Maria in 2017, sustained the second largest blackout in history: more than 3.4 billion hours of electricity were lost. But a few million people without electricity for days or weeks does not really constitute a world crisis.

Besides the supersized hurricanes, the summer of 2017 also witnessed record-breaking triple digit heat waves in northern California and wildfires in Greenland. In 2020, the western U.S. experienced the acute impact of climate change. At one point in the summer of 2020, 100 fires were burning in 13 states, torching more than 6 million acres, an area the size of New Jersey. In California, more than 3 million acres burned, killing more than 20 people. A fire north of Sacramento became the largest fire in state history, and six of California's 20 largest fires in history occurred in 2020. In Oregon, 1 million acres burned causing 500,000 people to be evacuated. Oregon had never seen the likes of this before. Entire towns disappeared. The following season (2021) the extent of wildfires in California broke the record of the previous year.

The increasing drought conditions, dry weather, and hot winds means this is not a onetime occurrence. It is the bellwether of the future.

Ignoring the warming of the planet and the inevitable consequences of that warming is an attitude limited to a minority in the U.S. Most everywhere else, the stark reality of what is going on has sunk in. The Paris Climate Agreement is based on the realization that the climate's threat to humankind is real, that its deleterious effects are already being seen, that they are becoming and will continue to become more and more horrific, and that this emergency requires a collective response from all the nations of the world.

Former Vice President Gore clearly expresses his frustration and disappointment that we have not made greater progress in reducing carbon emissions and stemming the planet's temperature rise over the course of the last 15 years. He should not be so hard on himself. He played an incredibly important role in drawing the world's attention to this situation. He may have just got the solution wrong.

It would be helpful if there were a reliable model to enable us to determine how much reduced carbon emissions on Earth will reduce the ambient CO_2 levels in the upper atmosphere. But there is not. It would also be helpful if there were some ways to know at what CO_2 level we can expect to see a slowdown, or much better, a reversal, of the current global warming trend. But there is not one of those either.

But these things may not matter. We may never know the exact degree that human-caused action is responsible for the warming trend. But what we do know is the globe is warming, the glaciers are melting, and we as a civilization have very little time to prepare ourselves to face the potential calamity of rising seas.

U.S. Citizens Only

This section is majorly addressed to U.S. citizens. There is no need to preach to an American about the world being full of division and distrust. We see it every day on the TV and in just about every aspect of our politics. But this is not just true in the U.S. There are struggles and stresses, disagreements and dichotomies, on every continent.

That makes it all the more amazing that almost 200 nations and territories (more than the total membership of the United Nations) came together and signed an agreement to work together to combat climate change and reduce global warming.[7] These nations span the entire range of political systems, religions, and economic models. You have dictators and democracies, working shoulder to shoulder with communists and monarchies. You have capitalists and socialists, working hand in hand with all the economic shades in between. What they are looking at and what they are failing to look at are the subjects of this book. But this section is dedicated to my American brothers and sisters, because only here can be found a substantial number of people who seriously contend that climate change is either irrelevant, nonexistent, or if it exists, not worth trying to solve. Climate change debates in the U.S. are usually divided along the following lines:

---

[7]Paris Climate Accords

Table 1 U.S. climate change debate: for and against

	For	Against
1	The Earth is heating up.	The Earth may or may not be heating up.
2	Increasing temperatures are already causing rising seas, flooding, wildfires, and ever-increasing, more destructive super storms.	If the world is heating up, it could be a normal warming cycle such as has occurred after the last six Ice Ages.
3	The warming trend is largely due to human-caused burning of fossil fuels.	The science linking global warming to the burning of fossil fuels is fallacious (or at least questionable).
4	Switching to wind and solar energies would reverse this trend and save the planet.	Three reasons suggest there is no urgency to take immediate action: (1) Whether humans caused it or not is uncertain. (2) The remedial actions proposed by international forums are hugely expensive, benefit the poor at the expense of the rich, are thus bad for business, and would likely be ruinous to the world economy. (3) If God or Nature is actually responsible, then there is nothing that can be done anyway.

With respect, the analyses and conclusions in both columns are inadequate. Blaming either God, the Devil, or Mother Nature is tantamount to diving head first into a hot oven to avoid having to watch your family engulfed by the onrushing seas. On the other side, blaming humans while betting all your kilowatts on a single solar/wind basket may not be the most provident or efficacious of actions.

In case you were not sure before, this book will show that the Earth is indeed warming up. As to whether this is human-caused or a natural cycle of nature, we do not offer an opinion; because even if it is not human caused, it is certain that the consequences of this warming, already peeking out at us from around the corner of the nearest melting glacier, will be devastating and only "human-caused action" will be able to mitigate those consequences.

Now that the U.S. has rejoined the Paris Climate Accords, we can acknowledge that all peoples on Earth are in agreement that it will take a concerted global effort to stand up to the challenges of global warming.

Chapter 2

Fossil Fuels' Double Duty: Warming and Poisoning[8]

It is all well and good to say "let's get rid of fossil fuels" but one cannot forget that the energy supplied by the burning of coal, gas, and oil underpins almost every one of society's comforts and conveniences. In countries where fossil fuels are the single most valuable natural resource (such as the U.S.), it is no surprise that attempts to curb CO_2 emissions are met with opposition. Fossil fuels are entrenched by huge corporations formed to find, extract from the Earth, refine, transport, and distribute them widely to consumers. Every person in the U.S. and around the world continuously benefits from the energy they create. In advanced economies, this amounts to a 24//7 availability of lighting, heating, transportation, manufacturing, medical care, cell phone charging, Internet communication and business, and so much more.

An equally powerful force driving greater use of fossil fuels comes from emerging economies craving more and more energy. The billions of people of China, India, Pakistan, and Indonesia (to name a few) have seen the advantages of cheap sources of energy and want more of those comforts for themselves. And as quickly as possible.

[8]I hate to use a word like poisoning that is so final and absolute, but once you have read this chapter, I think you will agree that what is occurring is Poison with a Capital P.

Energy Miracles: The Global Warming Backup Plan
H. B. Glushakow

ISBN 978-981-4968-18-8 (Hardcover), 978-1-003-28442-0 (eBook)
www.jennystanford.com

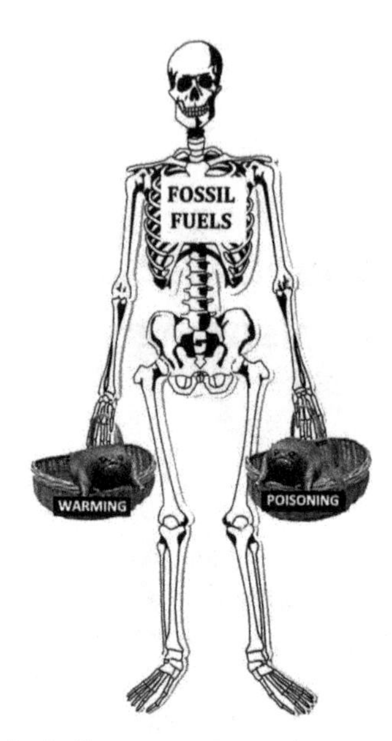

Figure 9 Fossil fuels dual effects: warming and poisoning.

These fuels appear to be easy to obtain and relatively cheap. To keep those prices cheap, governments (including the U.S. government) provide fossil fuel companies with huge subsidies and tax credits, and most ignominiously, ignore the real cost of the damage they wreak. But when 1 billion people are put in danger of being thrown out of their homes and workplaces on account of melting glaciers and rising seas, and when 95% of scientists believe this situation is being caused (or at least exacerbated) by the carbon emissions resulting from burning coal, oil, and natural gas, it is time for us all to sit up and pay attention.

If a UFO landed on Earth whose passengers wanted to keep the planet capable of sustaining life, the first thing they would put their attention on is finding a better fuel or some other energy source that would not burn chemical-fire fuels that smoke and throw soot and poison gases into the atmosphere. They would figure, rightly, that until they have and are using a better energy source, it would be useless to try and do anything else to salvage the planet. To do any real building or properly feed and water our populations, we need more fuel than is currently available—and fuel that does not pollute.

It is not an easy task to separate the actions and consequences of fossil fuel's warming and poisoning because there is so much overlap.

Those same deadly particles and chemicals that the burning of fossil fuels continuously dumps into the air, besides likely contributing to the warming of the planet, are at the same time causing many other extremely dangerous and deadly conditions across the world. Most would agree that the poisoning process has become "too much" when, according to the World Health Organization, one-third of the people of Earth have no access to clean drinking water and 90% of humanity must breath poor quality air. Not to mention the wars.

In this chapter, we look at these factors and the effects on the planet of each of them.

Warming the Land and the Sea: How Much Warming Is Too Much?

At least five studies have analyzed peer-reviewed climate science studies published between 1993 and 2013. Between 92% and 97% of the authors of the 12,000 surveyed papers concluded that human activity in the form of burning fossil fuels "is very likely" to be playing a major role in the pronounced and dangerous world temperature rise. For 5 years, the Intergovernmental Panel on Climate Change (IPCC) hedged its conclusions about the effect of human-caused actions on the climate. But with improved knowledge of climate processes and additional evidence, its last report, issued in August 2021, concluded that "it is unequivocal that human influence has warmed the atmosphere, ocean, and land." Three major energy sources (coal, gas, and oil), when burned, throw tens of billions of tons of harmful particles and chemicals into the air creating dangerous levels of CO_2, SO_2, NO, CFCs, PMs (particulate matter or particle pollution), and methane. Almost all scientists believe that these particles are causing the rise in temperature of the Earth's landmasses and seas.

Warming occurs when rays from the sun heat the Earth. The Earth absorbs some of that heat but reflects a large portion of it back into the atmosphere, which traps it. Ideally this process ensures temperatures remain within a range that can sustain human, plant, and animal life. The NASA diagram in Fig. 10 shows that of the total amount of heat from the sun hitting Earth, 29% is immediately reflected back into space with no impact on global warming; 48% is absorbed into the Earth or oceans; and 23% is absorbed within the atmosphere by atmospheric gases, dust, and other particles. Carbon emissions from the burning of coal, oil, and natural gas tend to thicken the atmosphere, thus trapping more heat. Carbon dioxide (CO_2) makes up the vast majority of these fossil fuel emissions.

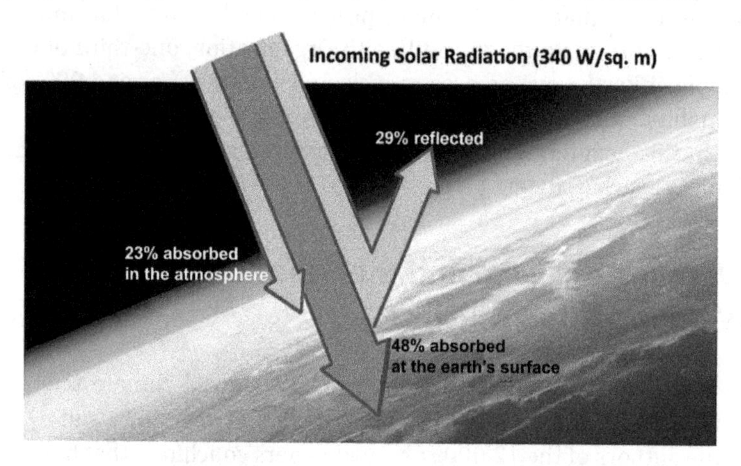

Figure 10 What happens to the sunlight that hits the Earth?

Today it is posited that our problem lies with the 23% segment of heat absorbed by the atmosphere. Too much of the sun's heat coming off the Earth is being trapped inside the atmosphere, causing a planet-wide heating process that warms the oceans and melts the glaciers.

The danger is not going to come from you waking up one day and finding the streets outside your home are hot enough to fry eggs on. A temperature rise of only 2 to 3°C is enough to melt the glaciers. We are talking about a steady temperature rise that will sooner than later have Earth-shattering effects. Here are some facts that cause concern:

1. The 10 hottest years on record have all occurred after 2005, including every year between 2014 and 2020.
2. According to NASA's Goddard Institute, the average global temperature rose only 0.07°C (0.13°F) per decade since 1880, but the rate of heating has more than doubled since 1980.
3. The rate of glacial ice melting has also doubled since 1980, because the temperature rises are not uniform across the Earth's surface. Greater temperature increases occur over

water than land, and the strongest warming is happening in the Arctic during its cool seasons. In many of these regions, warming has already surpassed 1.5°C above the 1900 level.

4. Per the IPCC, sea levels rose by 170 mm last century, but are now rising at double that rate (more than 3 mm per year).
5. If these trends continue, within a few decades, melting land ice and glaciers will add enough water to the oceans to raise sea levels high enough to flood major coastal cities on every continent (along with many islands and in some cases, entire countries). Studies vary over exactly how much 1°C will impact sea-level rise, but estimates of 2.3 m of rise per degree Centigrade seem to be the mean.

When residents of Mumbai, Shanghai, Miami, or Lisbon wake up to find their homes flooded, that is definitely "too much" warming.

Poisoning the Water, the Air, and the Global Social Environment

Fossil fuels (oil, coal, and natural gas) spearheaded much innovation in the world. Without them there would be no industry, no transportation, no personal computers, and no Internet. Even today, we rely on fossil fuels to meet the largest part of our energy needs. Depending on which statistics you believe, coal, oil, and gas provide between 79% and 89% of the world's yearly energy consumption. Figure 11 is derived from International Energy Agency and BP latest available data.

Meanwhile, there are several major climate crimes going on in plain sight all over the world that are not directly related to global warming. The planet is being poisoned by fossil fuels, and millions of people are dying for the lack of potable drinking water. We shall examine how these things are related and ultimately trace back to the principal cause common to them both: the lack of an adequate and clean energy source.

A favorite comment from those who do not consider global warming a clear and present threat is that "The Earth can handle it." This may have been the case a few hundred years ago, but the Earth is not doing so well here in the 21st century.

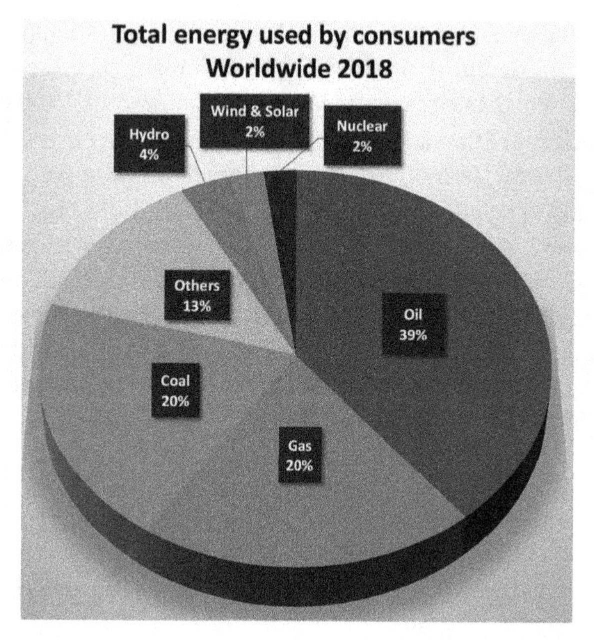

Figure 11 Total world energy consumption by source of fuel.

The Earth's air

It was once the case that you would wake up in the morning, open the window, and take in a deep breath of fresh air. Today, not many people are inclined to do this, including those living in the U.S. Instead, many resort to checking their smartphones for the daily air quality index.

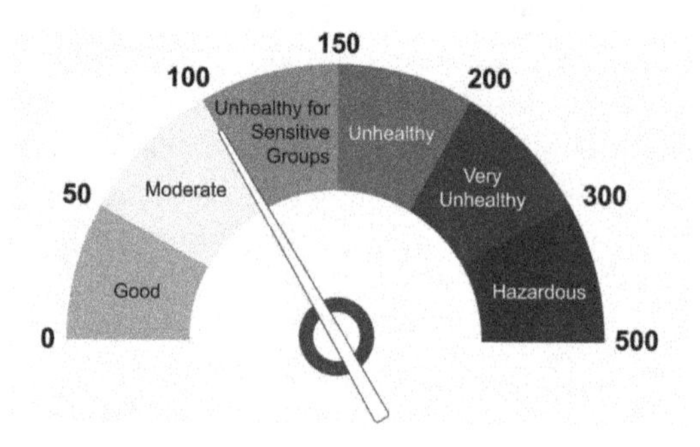

Figure 12 Index of air quality levels.

The main sources of air pollution include cars, buses, planes, trucks, and trains. Also, electric power plants, oil refineries, industrial plants, and factories are major sources. In today's world, fossil fuel combustion is the biggest contributor to air pollution, with the leading culprit being the fuel combustion from motor vehicles. The sulfur dioxide emitted from the combustion of fossil fuels such as coal, petroleum, and natural gas is the major cause of this air pollution.

During the 1960s in New York City, it was said in all seriousness that those who jogged daily along its streets were decreasing their life span by one day per hour of jogging. Yes, the air was *that* bad. Yet at this moment, half a century later, the American Lung Association warns that almost half of Americans—141 million people—are inhaling unsafe air.[9] This is an increase of more than 7.2 million people from the previous year, with eight cities reporting their highest number of days with dangerous ozone and particulate pollution level spikes since air quality was first measured 20 years ago. Figure 13, derived from NASA satellite data, shows the worst polluted locations in the U.S.

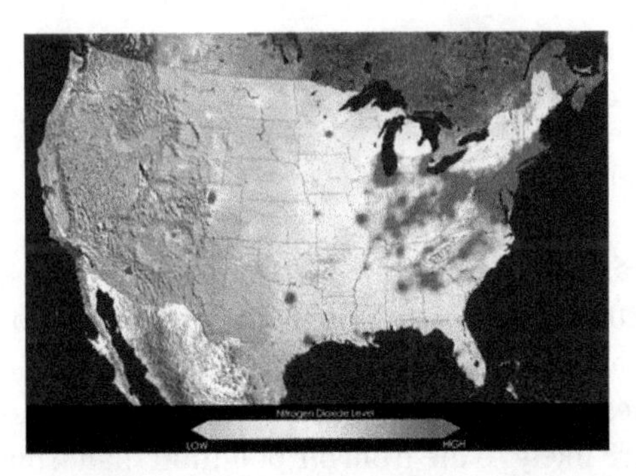

Figure 13 Worst polluted areas in the U.S. Nitrogen dioxide primarily gets in the air from burning fossil fuels. Examples: emissions from cars, trucks and buses, power plants, and off-road equipment.

In the good old days before California acted to handle its poisonous atmosphere, every time I flew into Los Angeles, I would experience 24 h of headaches and dizziness until the body adjusted. Since then, the strict emissions standards of California's Clean Air Act have made California cars and fuels the cleanest in the world.

[9]American Lung Association 2020 *State of the Air* report

But it is not enough. Los Angeles remains the city with the worst ozone pollution[10] in the U.S. with nearby Bakersfield, the number one worst polluted city by reason of particle pollution. Today, the states with the worst air quality include: Utah, Georgia, Ohio, West Virginia, Indiana, Tennessee, Colorado, Alabama, Maryland, and North Carolina.

The countries with the overall worst air pollution are all in Asia: Afghanistan, Pakistan, Mongolia, Bangladesh, and India (Fig. 14).

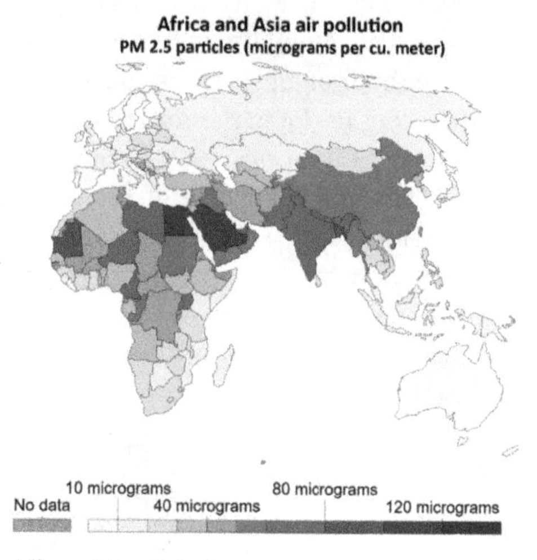

Figure 14 World's worst polluted areas.

Afghanistan, the country with the fourth most polluted air pollution in 2020, has been the site of bloody wars for over 40 years, long before the U.S.-led invasion in response to the September 11 attacks. As destructive as those wars have been to the country, an Afghan is more likely to die from air pollution than from war.

Pakistan was the second most polluted country in 2020, and 22% of Pakistani deaths are caused by air pollution according to a study in the medical journal *Lancet*.

Twenty-two of the world's most polluted 30 cities are in India, making it the fifth most polluted country in the world. The world's most polluted capital is New Delhi, with an average air pollution level more than three times higher than EPA-mandated safe levels. I

[10]Ozone is produced as a by-product of the internal combustion engines of cars and trucks and power plants.

was there not long ago when the air was so bad that breathing it for a few hours was equivalent to smoking 50 cigarettes a day.

The first time I visited China in 1992, when traveling to all parts of the country over a 3-week period, the sun never came out from behind the haze. I never saw it. Fifteen years later, Beijing's air quality was still giving China a severe black eye. Without firm promises of handling it, the People's Republic was not going to be allowed to host the 2008 Olympics. Based on China's promise to the International Olympic Committee that it would be handled, the Olympics came to Beijing. And the smog vanished. It was not magic: Vehicular traffic was severely regulated within the city and all industry was shut down. (It could not have happened in the winter when coal and gas-fired furnaces provide heat to Beijing's residents.) The Olympics left, and the smog returned. People began to grumble. Politicians were removed for failing to make good on promises to clean up the air.

Then one day the Asia Pacific Economic Council (APEC) came to town. Amazingly, the sky again turned a deep blue the entire week of the meeting, which prompted some critic to come up with a new color: *APEC blue.* This became the favorite color of those seeking reasons to criticize government leaders, and it went viral on Chinese social media. Scant attention was paid to the huge investment in energy and resources that had been put toward cleaning up the environment; or the fact that Chinese investment in wind and solar and its consumption of alternative energy now exceeds that of the U.S.; or the unprecedented tens of millions of trees that have been added to China's Great Green Wall—in the last 5 years China has increased its forested lands by over 20%, bringing the total to over 500 million acres; or the Beijing regulations forcing dirty industrial furnaces to vacate the city; or the vehicular traffic restrictions, investments into natural gas engines and bicycle sharing programs. But as is true everywhere, he who cannot do, can at least criticize.

And so, more than a decade after the 2008 Olympics, as I sat writing this in a small apartment overlooking Chaoyang Park in Beijing, for the past week the sky looked like this outside my window. I looked in vain to see if some fancy meeting was being held or maybe some important President or official had come to town. Nope. There was just the blue of the sky, as taken with my iPhone and un-Photoshopped.[11] This is only mentioned to make the point that humans can yet prevail over the environment.

Figure 15 APEC Blue in Beijing.

All over the world, pollution from the burning of fossil fuels is a major contributor to health conditions, including asthma, lung cancer, Alzheimer's and Parkinson's diseases, as well as heart disease. It is even said to worsen Covid-19 symptoms. As long as the planet continues its love affair with fossil fuels, air pollution will continue to worsen.

The Earth's water: drought is the new normal

"Water, water, everywhere, Nor any drop to drink"
Coleridge, The Rime of the Ancient Mariner (1834)

At first glance, there appears to be an abundance of water on our planet, with 70% of the planet's surface being covered with it. But the fresh water we need to drink, wash, grow our food, look after our livestock, and power our industry comprises only three-fourths of 1% of Earth's huge reservoir of water. Today, according to the United Nations, there are 2.2 billion people without access to safely managed drinking water. That is one-third of the world's population

[11]The smog returned a week later. But not as bad as before. The US Embassy's air monitoring system reports that Beijing has reduced the level of dangerous 2.5-micron particles by 20–30%.

lacking safe drinking water. They have no readily available water at all, or water contaminated with chemicals or feces.

Water is an area where there is the largest overlap between the warming and poisoning effects of fossil fuels.

Consider America's water. The United States of America: the most advanced country in the world with the highest standard of living. Between 40% and 50% of America's rivers and lakes are too polluted for fishing, swimming, or aquatic life. In the U.S., it is no longer safe to casually drink from or bathe in a stream or lake. Every one of the 50 U.S. states has consumption advisories to protect people from the health risks of eating fish that are caught in the local contaminated waters. In the world's largest and most advanced economy, you cannot safely drink the water. There is an old saying about fish that needs updating: "Give a man a fish and he'll be fed for a day; teach him how to fish and he'll die of mercury poisoning in just a few years." The biggest source of mercury in our oceans is the burning of coal, which releases 160 tons of mercury a year into the U.S. alone. The U.S. Geological Survey estimates that fossil fuel emissions have doubled the level of mercury in the atmosphere in the past 150 years. When this is washed into rivers and oceans, the fish get contaminated. At the same time, global warming is making the U.S. southwest ever hotter and drier.

Most people know the Colorado River because of the famous Hoover Dam on the Arizona–Nevada border that secures the water and provides electricity for 40 million people in the western U.S. The river also irrigates over 4 million acres of agricultural land. For 6 million years, the Colorado has flowed 1450 miles from its headwaters in Wyoming, Colorado, and New Mexico, down across what is now the international border into Mexico, and from there southwest into the Gulf of Mexico. Now due to 20 years of drought, the river slows to a trickle and dries up long before it reaches the sea.

Global warming directly impacts the Colorado River's water flow as increased temperatures have been shrinking the Rocky Mountain snowpack, a major source of water for the river. The dwindling volume of the river has lowered the water level 139 vertical feet, but that is not its only problem. Human overuse and environmental issues (such as agricultural overdraw) have increased the salinity of the river to levels that exceed the EPA's threshold of 500 mg/L for drinking water. In May 2019, seven states (including Arizona and California) plus Mexico signed the Colorado River Drought Contingency Plan (DCP). This agreement seeks to protect the water levels at the hundreds of hydroelectric dams within the Colorado River system[12] to ensure they can continue to generate electricity. The agreement provided for the enactment of severe water rationing in the event of worsening water conditions. In January 2021, following two more years of declining water levels and increasingly bleak forecasts for the Colorado River, DCP water restrictions were activated for the first time. A 2021 study from the Center for Colorado River Studies at Utah State University begins with the statement: "Our ability to sustainably manage the Colorado River is clearly in doubt." The report warns that the planning and water management techniques in current use are "unlikely to meet the challenges of the future."

Figure 16 Hoover Dam and Colorado River, taken by Ansel Adams. Public Domain.

[12]There are 15 on the main channel and hundreds more on its tributaries, but over 80% of the river's hydropower capacity comes from just two of them: the Hoover Dam and the Glen Canyon Dam.

As mentioned earlier, all 50 states in the U.S. have issued warnings and regulations to deal with unsafe water in the country's rivers and streams. The bottled water market in the U.S. is the largest in the world with 15 billion gallons sold in 2020. Americans pay over 50 billion dollars a year for drinkable water. And the situation gets worse as you move south in the Americas. In Mexico, nearly three-quarters of the population resort to drinking packaged or bottled water. In the Ecuadorian and Peruvian Amazon, indigenous people such as the Achuar are routinely confronted with oil spills in rivers and runoffs into lakes and forests. Pipelines carelessly shoved through traditional lands cause daily pollution events in the form of oil fires, gas and oil leaks, and waste dumping. This has resulted in high cancer rates and undrinkable water. Half the population of Haiti lacks access to clean water.

And of course, that is just America. In over half the world, it has been many decades since folks could safely drink water from their rivers or lakes. Almost all of Africa and the Middle East, much of Asia, and most of South America are experiencing severe water shortages.

In an Iraq teetering on the edge of stability, water shortages are threatening to topple what order remains. Deteriorating water quality on the Tigris and Euphrates, two rivers on which Iraq is heavily dependent for farming and drinking water, results in part from dams in Turkey, Syria, and Iran. With freshwater levels in those rivers reduced by the dams upstream, a vacuum occurs resulting in seawater from the Southern Persian Gulf backing up into canals and streams with devastating impacts for farms in the area. Over 2 million residents of the city of Basra, Iraq, have seen their drinking water contaminated. Basra, in the south of the country, was once known as the Venice of the East, for its abundance of water. Now drought coupled with water mismanagement has led to severe shortages and mass protests.

In the northern Iraqi Kurdistan region, there are brewing tensions over water between the Kurds and the Iraqis, as well as between the Kurds, Iranians, and Turks, who share borders and rivers.

Jordan, Lebanon, and Yemen are all on the brink of a water crisis exacerbated by a combination of climate change and government failures. Yemen is undergoing what the United Nations calls the worst humanitarian crisis in the world. The ongoing war and power struggle have been tied to drought and consequential food insecurity. Over half the population does not have access to a clean water source.

In Afghanistan, clean drinking water is available to less than 13% of the country.

Africa is the continent hardest hit by its water crisis. More than one-fourth of its people spend between 30 min and 6 hours walking an average of 3.75 miles just to collect enough water for the day. In many communities, women and children spend up to 60% of their day making the walk to collect water. In rural parts of the Congo, only 20% of the people have any access to nearby water. In Sudan, after 5 years of civil war left 400,000 people dead and another 4 million displaced, 80% of the country lacks access to clean water.

Pollution from the petroleum industry affects many countries in Africa, but Nigeria may be the worst. The Niger Delta (home to 20 million people and 40 different ethnic groups) comprises 7.5% of Nigeria's total landmass. The petroleum industry has made this area one of the most polluted areas in the world. For over 50 years, there have been literally thousands of oil spills every year, the result of pipeline and tanker accidents, sabotage, and old and corroded infrastructure. Many millions of barrels of oil have been spilled into the Niger Delta directly contaminating its water resources and causing many health issues and destruction of crops and fish. In Ethiopia, 50 million people lack access to potable water.

In Asia, most of Indonesia's population is now also addicted to bottled water due to the continuous pollution of its water sources from domestic and industrial wastes. Indonesia is home to the Citarum River, labeled the world's most polluted river by the World Bank. Citarum, which supports over 30 million people, is now dense and smelly with a dirty brown color except when it changes to black, blue, or red as a result of the 300 tons of toxic waste dumped into it every day. Lead levels are 1000 times higher than U.S. maximum levels for drinking water. In Cambodia, 84% of the population lacks access to clean, safe drinking water. In India, over 20% of communicable diseases can be traced back to unsafe water. Such examples are increasing all over the world.

Over the past three decades, bottled water has become the world's fastest growing drink market, valued at over $250 billion. Water sales are 130 times greater than they were in 1980. But that water is only for the relatively rich.

How fossil fuels are directly robbing us of needed water

Water typifies many of our global problems: In the midst of plenty, we somehow have scarcity. As with anything scarce, there is a push/pull of priorities. The greatest domestic users of water in the U.S. are the agriculture industry and thermoelectric power industry.

Total withdrawals for thermoelectric power account for 41% of total water withdrawals, 34% of total freshwater withdrawals, and 48% of fresh surface-water withdrawals for all uses. Stated another way, thermoelectric power withdraws about 200 billion gallons every day from the country's lakes, rivers, and underground water sources. The mining and public utility industries with their thermoelectric power plants withdraw 75 trillion gallons of water annually to produce and burn the 1 billion tons of coal America uses every year to produce electricity.[13]

Billions of gallons of water are used in the power plants themselves in the process of converting water into high-pressure steam to drive the turbines, and then cooling that water sufficiently so it can be disposed of. In the mining of coal, the debris from mountaintop removal is often pushed into streams further depleting freshwater supplies. In coal mining operations, large quantities of water are used daily to cool and lubricate mining machinery and wash the mountains of coal loaded into trucks and trains, in order to reduce airborne particulates, and to suppress underground coal dust that could otherwise ignite. Cooling water is the largest source of water withdrawal in developed countries (accounting for up to 50% of national water withdrawals). In less affluent countries, this can fall to 20% of total water withdrawals. But in any country, a single 500 MW coal-fired electrical utility uses up to 20 million gallons of water per year just to cool its equipment.

And of course, there is fracking. The term is short for *hydraulic fracturing*. After a deep well is dug in the Earth, large quantities of water mixed with chemicals are forced down the well at very high pressure—high enough to fracture the rocks below which then release oil and gas from shale rock. Fracking uses huge amounts of water—up to 100 times more than traditional extraction methods. The fact that the water is mixed with poisonous chemicals endangers

[13]U.S. Geological Survey

the local environment when the wastewater that comes out of the well's mouth mixes with local groundwater.

Global warming and the Earth's social stability

> *"We must concentrate not merely on the negative expulsion of war but the positive affirmation of peace."*
>
> —Martin Luther King, Jr.

When global warming results in conditions of severe water shortages, a society can go badly off the rails with immediate and surprisingly destructive consequences. Such is the case with Syria, causing the largest displacement and refugee crisis of our time. At this writing, the country has been engaged in over a decade of violent civil war, a conflict that has so far resulted in half a million casualties (3% of all Syrians). Between five and six million Syrians have fled their homeland, seeking asylum in other countries. Another six million are displaced within Syria.

There is an overly simplified story of how the Arab Spring was caused by the spirit of democracy sweeping through the Middle East and Northern Africa in 2011 resulting in popular uprisings toppling authoritarian regimes. But one of the Arab Spring's main triggers was the spike in food prices in Egypt and Tunisia in 2011. Russia had been the main source of wheat imports for those two countries, but a drought in Russia 6 months earlier forced Moscow to cease all wheat exports for the first time ever.

The effects were worse in Syria, where there had already been widespread social unrest and demands to oust the government long before the Arab Spring. The Arab Spring's main impact on Syria was an almost total collapse of government services except for governmental violence toward its own citizens, which was escalated. The thing is, there was an earlier beginning to this incident. How come the country was so unstable that its social fabric could break down so quickly and so thoroughly? The answer is: **lack of water**.[14]

[14]See "Report on National Security Implications of Climate-Related Risks and a Changing Climate," a U.S. Department of Defense report on global climate change, September 30, 2015.

Figure 17 The 2009 drought preceding the Arab Spring.

From 2006 to 2010, before the civil war, and before the Arab Spring, Syria had endured its worst drought in at least 900 years. Before the drought, agriculture accounted for one-fourth of the country's GDP. The drought wiped out all that production, and when farmers were producing zero livestock, zero wheat, and zero rice, foodstuffs had to be imported. But these imports were so expensive that most people could not afford them. Two million people were displaced, and 60% of farms and 80% of livestock were destroyed. Starvation and illness reigned. All that before any conflict broke out. The stress put on Syrian society by the water shortage was the major cause of the subsequent events of the Arab Spring and the civil war. At this writing, the country is in a terrible mess. Twelve million people inside Syria need humanitarian assistance, half of them children. But the "Syrian Crisis" has also had a major impact far beyond its own borders.

The situation inside Syria became so toxic that almost anyone who could flee, did flee.

Six million Syrians (30% of the population) fled their homeland, appearing on the doorsteps of other countries. Some were granted asylum; others were shoved into refugee camps. Syrians accounted for the largest percentage of the 2.5 million refugees who have fled to the European Union since 2010. These refugees became both a humanitarian and political problem to the countries concerned and exerted major stress on the European Union itself. The 50,000 refugees in Greece were one of the factors that almost led to Greece leaving the European Union. The political crisis over that migration and its effect on the European Union tipped the United Kingdom's Brexit vote by just enough to result in the British leaving the European Union. All, stemming from **lack of water**.

The solution generally offered to handle the water crisis is "Conserve your resources," but that is not much of a solution when the planet's immediate basic water requirements so greatly exceed the amount of available safe drinkable water. The only solution is to provide more water, and that is only feasible with a cleaner and more abundant source of energy.

There is technology available to render seawater drinkable. For example, reverse osmosis (RO) can filter out everything that is harmful in seawater (including the salt) and turn it into a potable drink. Its main drawback is that the technology is energy intensive, making the water it produces uneconomical. All that is lacking to instantly solve the world's drinking water crisis is an inexpensive source of electricity. Think of it: a billion thirsty people with sudden access to clean drinking water. A billion hungry people with sudden access to food because with the injection of clean water, the deserts could once more support agriculture.

Persian Gulf miracle

Water is an element that flows in many miracles. Famous water miracles include parting and then crossing the Red Sea. Also walking right across the Sea of Galilee, and summoning water out of a rock. Myths from numerous cultures recount a time in ages past when water destroyed the entire world. For centuries Europeans have frequented exclusive spas to "take the waters" in hopes of healing physical or mental ailments.

I have always thought those events to be third-class miracles at best. A first-class miracle should be something like creating, out of the blue, a bright shiny object with weight and mass that could be seen by others. And on my first trip to Dubai, I witnessed such a real modern-day water miracle.

Dubai is the capital of the United Arab Emirates (UAE), an arid country situated between Oman and Saudi Arabia on the western shore of the Persian Gulf. It has very little fresh water, yet it boasts the highest per capita consumption of water in the world. It is also the world's most modern and advanced city.

Figure 18 Dubai, world's most modern and advanced city.

Fifty-five years ago, before the discovery of oil, Dubai was little more than a bunch of camel herders sitting around some tents with a platoon of Brits cracking whips. In the time since then it has become the most modern advanced city in the world with over 200 skyscrapers, including the tallest building in the world. Its super-modern and efficient airport is the world's busiest, and all citizens are afforded free Internet.

Figure 19 Dubai Creek: 1964 and today.

You could call Dubai: Tomorrowland. It holds 220 awards in the Guinness Book of World Records, including world's tallest building, city with the most buildings over 300 m tall, fastest police car in service, highest tennis court, largest indoor ski resort, largest piece of art, largest gold chain... The list goes on and on.

In the times when Bedouins reigned supreme, water was priceless, with the oasis being the central point of the culture and of life. Without water none of the preceding wonders could have been accomplished. Yet they were. So how did they do it?

Without discounting the foresight and determination of its leaders, this miracle was accomplished by desalination plants such as the Jebal Ali plant (Fig. 20), which produces 564 million gallons (2.13 billion L) of pristine drinking water every day to meet the needs of Dubai's burgeoning population (and golf courses).

Figure 20 Dubai's Jebal Ali desalination plant.

Without water, the miracle that is Dubai could have never occurred. That water did not come without a price; 99% of Dubai's water comes from desalination—the process that produces drinking water from seawater. After Saudi Arabia, the UAE has the highest desalination capacity in the world. Most of its desalination plants use cogeneration multistage flash (MSF) technology or multiple-effect distillation (MED). Two plants use reverse osmosis (RO) technology. All of these technologies are energy intensive. The complete system comprises approximately 30 turbine generators running on diesel fuel and natural gas. These are major contributors to Dubai's skyrocketing CO_2 emissions.

Such emissions are poisonous to the environment and not sustainable. But what if you could provide unlimited amounts of cheap energy that does not require burning any fossil fuels and so produce zero emissions? That is where we are going in this book.

United Arab Emirates carbon dioxide emissions from fossil fuel 1970 to 2019 *(in million metric tons)*

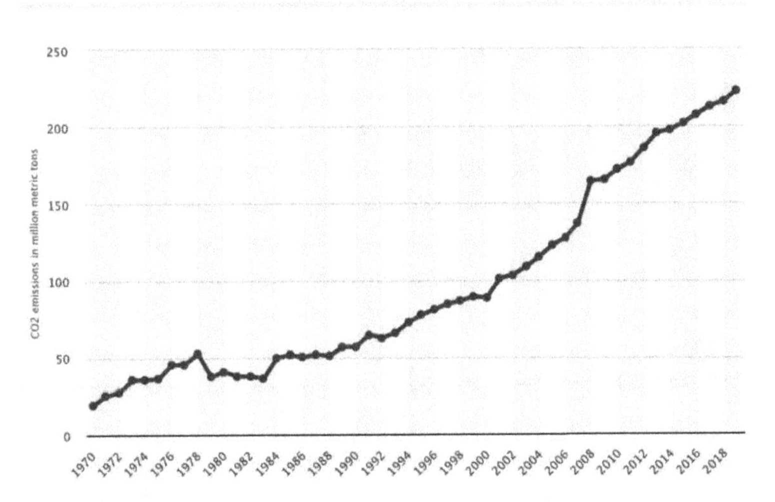

Figure 21 UAE's carbon dioxide emissions.

Chapter 3

The Valiant Effort to Replace Fossil Fuels

Between 2010 and 2020, global investment in renewable energy capacity exceeded 2.6 trillion U.S. dollars, and 1.3 trillion went to solar power. Another trillion went to wind energy. This investment accomplished a fourfold increase in global renewable energy capacity (if you exclude hydropower). In a single decade, wind and solar capacity increased from 414 GW to 1650 GW. As magnificent as that may seem, in that same 10-year period, global power sector carbon emissions rose 10%.

Figure 22 shows wind and solar worldwide consumption at the top right corner. The combined wind and solar production accounts for less than 2% of total energy and between 2% and 7% of electrical power usage (estimates vary). It is too bad that a mere chart cannot show the blood, sparks, and spittle expended by the wind and solar energy folks in carving out this small but solid toehold in the global energy market. It has taken 30 years of struggle with nonstop promotion, trillions of dollars of investment, and continued government subsidies to those two technologies. This was accomplished in a period when the world's economy was relatively strong and expanding. With the economic effects of the Covid-19 pandemic expected to stay with us for years and the world now arguably in a global recession, maintaining previous levels of investment and expansion in alternative energies is uncertain. These

Energy Miracles: The Global Warming Backup Plan
H. B. Glushakow

ISBN 978-981-4968-18-8 (Hardcover), 978-1-003-28442-0 (eBook)
www.jennystanford.com

technologies must continue to be supported, but they will need help if we are to fully get the job done of replacing fossil fuels.

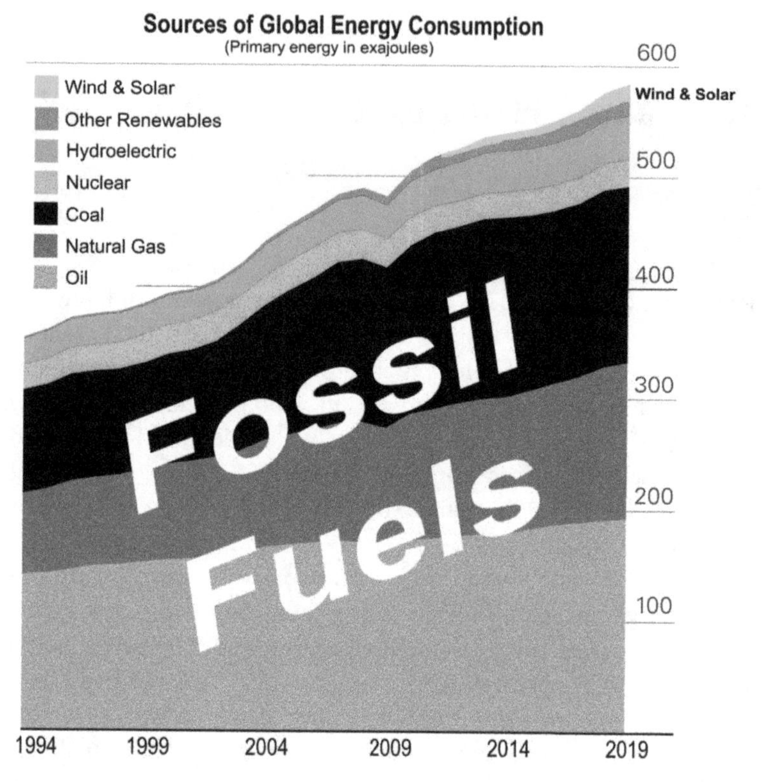

Figure 22 Global energy consumption by source (data from BP).

Transition to Renewables: Some Tough Truths

The planet really has no choice but to phase out fossil fuels, and the sooner the better. The thing is they must be replaced by an **alternative** energy source that is cheaper and just as reliable. Alternative energy sources these days include biofuel (fuel made from plant sources), wind, solar, and hydro (energy from water such as dams across running rivers). Some add nuclear power plants to that mix. When you look at it like that, renewables contribute 19% of our global energy consumption.[15] Not too shabby on the face of it.

[15]Renewable Energy Policy Network for the 21st Century

But when "investment in renewables" is promoted and discussed at international conferences, including the Paris Accords, discussions mostly center on wind and solar, which contribute only a small fraction of that 19% as can be seen in Fig. 23.

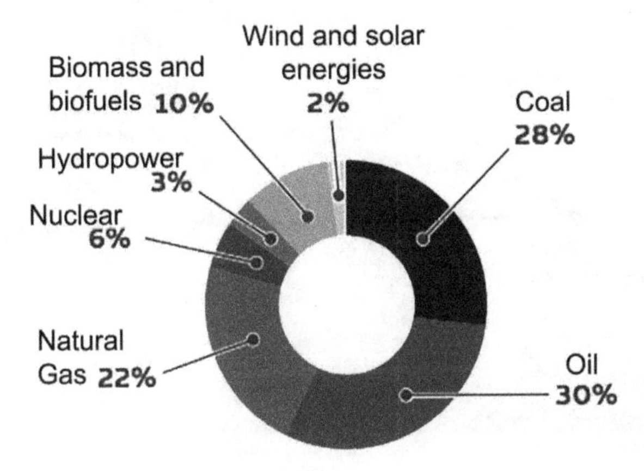

Figure 23 Renewable energy from wind and solar is less than 2% of total.

We know we need more energy produced to meet the most minimal needs of the planet, and that energy must not result in dangerous carbon emissions. And we know all forms of energy are not the same. A wind turbine can provide electricity to power the appliances in a home but could never power an airplane. A solar panel is relatively effective in climates near the equator, but its efficiency greatly declines the nearer it approaches the poles. Any way to rid ourselves of the carbon emissions of cars and trucks is acceptable, but if you are using electricity from coal-fired turbines to charge the electric car batteries, you are taking three steps forward and two steps backward.

The International Energy Agency's "World Energy Outlook" reports that despite 25 years of mega attention to alternative energies and gigantic government subsidies, "Today's share of fossil fuels in the global mix, at 82%, is the same as it was 25 years ago." In other words, all the work gone into promoting alternative energies has thus far made no substantive inroad in improving the world's overall energy picture.

Figure 23 corroborates other sources that report the percentage of the Earth's energy needs that wind and solar are thus far able to supply. The world's energy consumption has more than tripled in the past 50 years. Figure 24 shows there is no sign of abatement in that trend. Experts forecast that it will grow another 50% by 2050. What sources of power will provide for that demand?

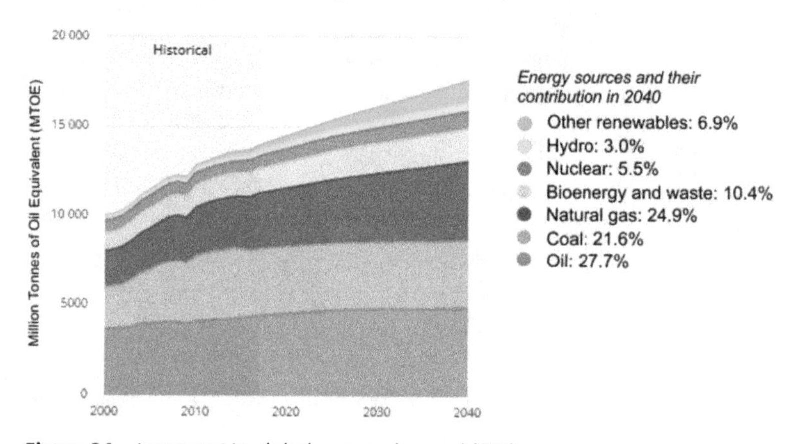

Figure 24 Increases in global energy demand (EIA).

A caution on renewable energy ratings

While we are evaluating renewable energy statistics, here is a caution about their ratings. Most statistics you see speak of "installed ratings" or "capacity." That would be the total ratings in watts of which the wind or solar installation is capable at any one point in time. Three wind turbines each rated at 500 MW would give a total installed rating of 1500 MW. Elementary. The problem is that the huge ratings you sometimes see on wind turbine generators often mask the true energy potential of the wind turbine. You rate a generator by its peak power output. If a diesel-powered generator is rated at 500 MW, you can easily compute how many megawatt hours it will generate in a year, by multiplying 500 (MW) by 24 (hours) × 365 (days). That is because a diesel generator can be counted on to run 24 hours a day every day of the year. It is steady, and it is dependable.

A wind turbine generator with the same "500 MW" rating on its label is hiding a critical fact. That 500 MW peak power can only occur within a small range of wind speeds and directions. Because wind energy is so intermittent and so variable, you will never get anywhere near the amount of power out of a wind turbine generator that you would from an equally rated diesel generator that runs 24 hours a day. Wind power varies as the wind speed increases or decreases or as the wind hits the turbine from different compass directions or horizontally or from above. No power at all is produced if the wind speed drops too low. And for safety reasons, wind turbines are programmed to turn themselves off when the wind speed gets too high. Depending on a wind turbine's location, the time of the day, the season, the existence of nearby structures, and other factors, you may only get 20–30% of the "rated" power over the course of a year, and it could be even less.

This holds true for solar installations as well. A solar plant rated at 500 MW can produce that level of power only under ideal sunny conditions. It will produce much less on cloudy days and nothing whatsoever during rainstorms or in the night hours. The true production of solar power is between 11% and 35% of its rated capacity reflecting available sunlight.

Wind energy disappointments

Go visit a wind park—any size, large or small; from Germany to Xinjiang Province in China; from Atlantic City to Bahrain to tropical Hainan Island. You will often find a large percentage of the turbines not turning. Sometimes an astonishingly high percentage: as much as 30%. Even on windy days.

There are many reasons for this. Lightning is one of the major ones. The electronics in a wind turbine are very susceptible to damage from the voltage surges created by lightning. Rarely are these the result of direct hits. More often these electric surges invade the turbine's control system by coupling onto power lines connecting the turbines to the grid. An even greater source of damaging transient surges is the wind turbine's generator itself. Continuous blade adjustments to compensate for changes in wind speed and direction create an almost continuous barrage of damaging surges that stress the wind turbine's control system and eventually cause a shutdown. Wind turbine manufacturers consider they are doing well if they can get by with only two shutdowns per wind turbine per year. Thus, in a country like China with 25,000 wind turbines, you are looking at 50,000 shutdowns per year. Once damaged and offline, these wind turbines are often difficult to access and repair. Wind turbines constructed offshore are even less easily accessed and even more susceptible to damage, not only from lightning, but also from the corrosive salts of the sea and the pounding of waves and storms.

Figure 25 Abandoned Hawaiian wind turbines after 18 years of service.

Wind turbine generators have another problem, sometimes called wind shadow. Wind does not flow steadily from a single direction. It will shift and pulse in completely random ways. The presence of nearby structures will severely interfere with the flow of wind to a turbine. It is fashionable to install as many wind turbines as will fit in a high wind area to "maximize the potential energy that can be generated," but installing wind turbines in close proximity, as done in many wind parks, is like installing a solar panel in the shade beneath a tree.

Wind shadow cannot be better seen than in the example of Pearl River Tower, in Guangzhou, China, completed in 2011, and considered to be one of the most energy-efficient super-tall buildings ever constructed. The building is owned by the Guangdong Tobacco

Corporation, and Guangdong Tobacco oversaw every step of its design and construction and ensured every floor included high-tech smoking rooms.

Figure 26 Wind shadow.

Hundreds of millions of dollars went into designing and constructing this 71-story building exotically sculpted to channel any wind hitting the building from any direction into four designated wind tunnels (two each on the 24th and 50th floors). It was predicted that this design would provide between two and eight times the amount of energy that could normally be obtained at the building's ambient wind conditions and produce 132 MWh per year. Beautifully designed Windside vertical axis wind turbines were installed in each of the four wind tunnels. So, what was the result?

Unfortunately, Guangdong Tobacco crowbarred the building into a narrow space in the wind shadow of a bunch of other skyscrapers. Figure 27 is a Google Earth view of the Pearl River Tower showing the nearby buildings interfering with incoming wind flows. The building has received various environmental awards, including LEED Platinum certification, and tourists can visit the wind turbine observation room to get a close-up view of one of the four 10-m-high turbines. But putting aside public relations, Guangdong Tobacco Corporation found it had wasted the money it had put into that sleek building design and wind turbine technology when the energy produced from wind at the Pearl River Tower turned out to be negligible. The only advantage obtained from those four wind tunnels ended up being some minor savings in construction costs when the reduced pressure differential they created, between the building's front and back sides, allowed the use of less steel and concrete.

Figure 27 Pearl River Tower in wind shadow (showing its four wind tunnels).

The Pearl River Tower was not the first failure at obtaining energy from building-integrated wind turbines. The Bahrain World Trade Center, completed in 2008, is a 50-story double-towered structure in the Kingdom of Bahrain.

It was the first skyscraper in the world to incorporate wind turbines into its design. On a recent trip to Bahrain, my hotel was just down the street from the WTC and I had the chance to pass it numerous times every day of my stay. Some of the time the wind was gusting up to 10 m/s, but none of the three wind turbines ever turned at any time during the 10 days of my trip.

Figure 28 Bahrain World Trade Center wind turbines.

Solar energy disappointments

Creative ways to boost solar power up above its 0.3% share of total global energy production have thus far been lackluster. Constructing roads out of solar panels was one promising idea. In 2019, after only 2 years of operation, the world's flagship solar road project in France was declared a failure and demolished. The surface had deteriorated beyond recognition, and it had produced less than half the power originally estimated, due to thunderstorms (which caused short circuits), tractors (whose weight crushed the panels), and rotting leaves, shade, dirt, and the cars themselves, which all interfered with the operation of the solar panels. Solar panels cannot operate when the sunlight needed to produce energy is blocked.

Figure 29 Normandy Road paved with solar panels.

Concentrated solar thermal (CST) plants are another attempt at broader application of solar technology. The project in Medicine Rock, Canada, used huge, curved steel panels to concentrate the sun's rays in the effort to improve efficiency. The heat generated would boil water to run a thermoelectric power plant that was to replace the existing gas-fired plant. After 5 years of operation, and an investment of $12 million, the project was abandoned in 2019 when it became evident that Canada was not such a good location for solar panels, even when concentrated.

India: a people crying out for more energy

India, the world's second most populous country, almost single-handedly scuttled the Paris Climate Accord. The country was set to construct hundreds of huge coal-fired power plants to support its economic growth. The country's industry required it, and its population (25% of whom had no access to electricity) demanded it. Indian leaders had no taste for slowing down their country's growth and were willing to accept the noxious atmosphere of Delhi (with the worst air quality in the world) and the 40 million people who are at risk of being displaced as rising ocean levels invade the country's coastal cities. For these reasons, India was not going to sign the Paris

Accord. But with its more than 300 days of sunshine per year, India has one of the best conditions for solar power in the world. Al Gore played an important role in getting them on board. The first step was to bring the country's leaders to the understanding that if the predicted damage from climate change occurred, India would be one of the most seriously affected countries. (Mumbai, India's financial capital, and Kolkata are two of the world's most endangered cities from rising sea levels.) Gore then helped broker a deal to transfer solar technology to India to enable them to manufacture their own solar panels. With the addition of the promise of financial assistance from the more affluent countries within the Paris Accord to help make its energy transition realistic, India became a staunch supporter and signed the agreement. The Indian leaders were already facing domestic criticism for their failure to handle the abominable Indian air quality, and embracing the Paris Accord became a win–win situation.

India, after signing the Paris Climate Accords, and with the World Bank supporting it, has run with the ball, now boasting the world's third fastest rate of solar growth. It installed the two largest photovoltaic power stations in the world (the 2245 MW Bhadia Solar Park in Rajasthan and the 2050 MW Pavagada Solar Park in Karnataka) and is now investing more in renewables than in coal. Despite these great achievements, the use of coal continues to grow in India by 10 MW per year. And consequently, CO_2 emissions also continue to grow. The reason is economic growth. With its 1.3 billion people, India is the world's third largest consumer of electricity. As a country, it is using eight times as much power as it did 50 years ago. But 240 million people still do not have legal electricity connections, and demand continues to grow as the hundreds of millions of people in rural and impoverished areas seek access to electrical power in their homes and workplaces.

India is now expected to reach less than 70,000 MW of the 100,000 MW new solar capacity target set for 2022. They started falling behind in 2019 (even before the Covid-19 pandemic) when several government agencies and one private electrical distributor cancelled 8000 MW of new projects. In India, coal produces 10 times more energy than solar, and it is far from certain whether solar will ever be able to fully replace it.

Irish disappointments

As we have seen in India, the best laid plans to reduce carbon emissions do not always succeed. Ireland is another example. In 2009, the European Union issued its Renewable Energy Directive. In compliance, Ireland set a target to produce 16% of all its energy needs from renewable energy sources by 2020. Between 2005 and 2018, Ireland's percentage of renewable energy grew from 1.3% to 7.2%, a significant increase. By 2019, Irish-installed wind power capacity had increased four times compared to its 1990 level (from 1027 MW to 4155 MW) and installed solar PV capacity increased 90 times from 0.4 MW to 36 MW. Despite those accomplishments during that same period, Ireland's total greenhouse gas emissions rose 10% to 60.7 million tons (up from 55.4 million tons in 1990). According to the Irish EPA, this rise was due to a booming economy and increases in transport and agricultural activity.

Germany: alternate energy's poster child

Countries who can invest should definitely invest in wind and solar. But to those already in economic difficulties, just telling them to do it may not be enough. They may look at the example of the wind power poster child: Germany. Germany has been called "the world's first major renewable energy economy." In 2020, 38% of its electrical power was generated from wind and solar sources and, for the first time, wind and solar together produced more electricity than the fossil fuels (Fig. 30). This was partly due to an overall decline in electricity use due to the Covid-19 pandemic, but regardless, it was a magnificent accomplishment, and for that reason, Germany is often pointed to as proof that wind and solar energy can be the primary power sources of the future.[16] Unfortunately, it is not so cut and dried.

[16]In 2020, the Covid pandemic reduced electricity consumption in Germany by 15%. At the same time, electricity from coal declined by 35% due to an increased price of CO_2 certificates. This shortfall was majorly replaced by natural gas. (Data from Fraunhofer Institute for Solar Energy Systems)

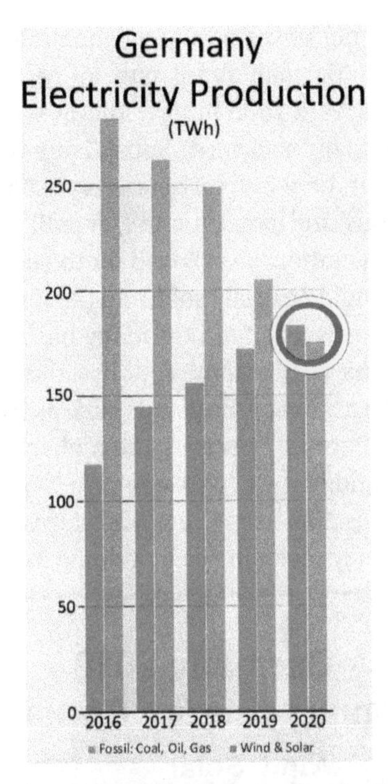

Figure 30 German electricity production. For the first time, wind and solar produced more electricity than fossil fuels in 2020.

In the first place, the above 38% figure applies to electricity only. It does not account for the heating sector of which 90% comes from heat boilers powered mostly by coal-fired generators. It also does not account for the transportation sector of which wind and solar accounted for only 5%.

So overall, wind and solar energy comprised less than 12% of total consumed energy in Germany in 2019 (Fig. 31).

As regards electricity from German hydroelectric plants, a few years ago electricity from hydropower far exceeded all other renewable energy sources, but this is no longer the case. Due to drought, the amount of electricity generated from hydroelectric power has plummeted 40%, from a high of 30 TWh in 2000 down to 18.4 TWh in 2020.

Good news is that the German government's "Energy Revolution" has cut carbon emissions almost 40% (from 1251 million tons in 1990 down to 858 million tons in 2018). But the program's one-trillion-dollar price tag poses a huge economic gamble. Part of its popularity lay in the German abhorrence of nuclear power plants, and who would not want to eliminate the possibility of a nuclear reactor blowing up in his backyard? Subsidizing German companies like Siemens (maker of wind turbines) was, likewise, a popular move—good for jobs and good for the overall corporate climate. But this "Energy Revolution" was based on the expectation that the generated energy would be easily sold, and the jury's still out on this. The cost of domestic electricity in Germany has risen by 30% since 2010. Next to Denmark, Germany now has the highest electricity prices in the world. Large multinationals such as BASF have cut back their operations in Germany because the higher price of electricity makes their operations less cost-effective. Without continuing government subsidies, the electrical power derived from Germany's Energy Revolution may make the cost of doing business in Germany prohibitive.[17]

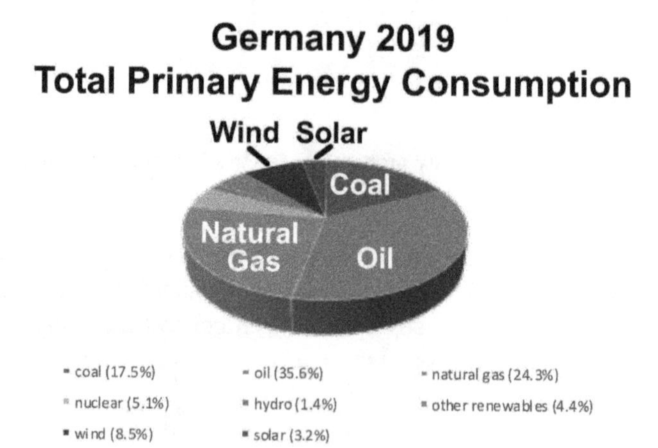

Figure 31 In 2019, wind and solar accounted for 11.7% of Germany's total energy needs. (Data extracted from BP Statistical Review of World Energy 2020)

Yes, we need to find alternatives for fossil fuels. And, yes, wind and solar power produce electricity without emitting CO_2. That is terrific. And for that reason, they should both be thoroughly supported. But at 11.7% of total energy usage, can we really claim

[17]Bundesministerium für Wirtschaft und Energie: Energiedaten: Gesamtausgabe. Stand: Okbober 2019.

these two "alternative" energies are winning the war against the fossil fuels in Germany?

China

Decades of rapid economic growth have dramatically expanded China's energy needs. In 2018, China's energy consumption accounted for one-quarter of the world's total. Even before the Paris Climate Accords, China had realized the need for green energy sources, and over the past decade, she has emerged as the global leader in both wind power and solar photovoltaic (PV) energy. Its renewable energy sources produce double the energy of the number two country, the U.S., and its renewable energy sector is growing faster than its fossil fuels sector. In 2018, China accounted for over 25% of the world's wind-energy generation and 66% of the world's solar production capacity.

On top of the success of its wind and solar programs, large-scale infrastructure investment has made hydroelectric power China's chief source of renewable energy production. The Three Gorges Dam, completed in 2012 at a cost of over $37 billion, is the largest hydroelectric dam in the world with a generation capacity of 22,500 MW. This dam generates 60% more electricity than the world's second largest hydropower dam, the Itaipu Dam in Brazil and Paraguay. In total, China has constructed four of the 10 largest energy-producing hydroelectric dams in the world. From 2000 to 2017, China more than quintupled its generation of hydroelectricity, from 220.2 billion kWh to 1145.5 billion kWh, making China the undisputed world leader in hydropower.

Domestically this translated to wind, solar, and hydroelectric together providing 11.9% of China's total energy consumption in 2018. The pie chart in Fig. 32 shows how this percentage of renewables is shared.

As can be seen in Fig. 33, China is far from independent of coal and oil for generating its power, accounting for over half the world's total coal consumption. Despite all its advances in wind, solar, and hydropower, its coal dependence keeps China at the top of the list of the world's CO_2 gas emitters.

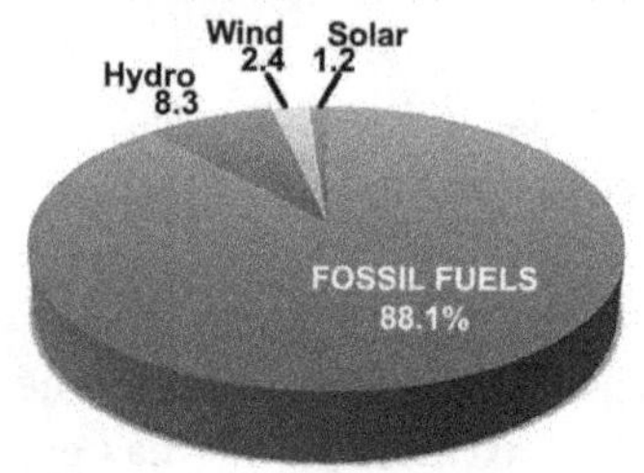

Figure 32 China's power generation by sectors.

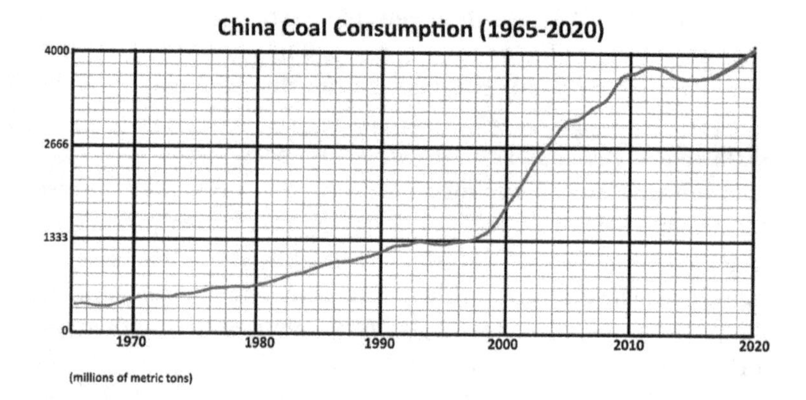

Figure 33 China coal consumption (data from BP Statistical Review of World Energy 2020).

More Energy Needed, Not Less: Some Consequences of Energy Scarcity

If we are only seeking to replace the **existing** uses of fossil fuel energy, we are on the wrong track. Electricity scarcity already poses a major threat to the world. Increased electricity demand is driven by the following:

1. Rapid population growth in concentrated urban areas; not just in the wealthy West, but all over the world.
2. The demand of rural and impoverished populations in developing countries to have electrical power in their homes and workplaces.

3. Addiction to "high-consumption" lifestyles dependent on advanced electrical appliances. This is no longer limited to a few developed countries. The billions of people in China, India, Pakistan, Indonesia, and Latin America have seen the lifestyle of cars, microwaves, and air conditioners, and want some of it for themselves.
4. Growing shortages of food and water—situations that foment continuing catastrophes that keep half of the world unstable.
5. The need to deal with the coming massive population movements and reconstruction as rising seas caused by planetary warming disrupt long-standing cultural and living patterns.

None of those five things should be news, but taken together they are evidence that our 19th-century sources of energy are inadequate and unsustainable.

Worldwide blackouts

The other day I received a check in the mail for $500. My power utility company was reimbursing me for purchasing an "energy efficient" hot water heater. They were paying me for using *less* power. But why would they be doing this when they obviously make more money when I use more electricity? The simple answer is that they do it to prevent blackouts. Blackouts (or power outages) are the results of breakdowns in the network of electricity grids. During peak demand periods, when the electrical supply companies cannot meet the need, grids become overloaded, and blackouts occur. Such blackouts are becoming more and more frequent in the U.S., due to increased needs, the proliferation of severe weather events, and the fragile national grids that are unable to keep up with a rocketing energy demand.

Every year, millions of people around the world experience major electricity blackouts, but the country that has endured more blackouts than any other industrialized nation is the U.S. Over the last decade, the number of power failures affecting over 50,000 Americans has more than doubled, according to Federal data.

A U.S. government program seeking to reduce electrical usage by encouraging the use of energy-saving technologies cuts the wholesale price of electricity to electrical suppliers in 24 states. Those utility companies pass on some of that savings to their customers with "energy saving" rebates such as the check sent to me.

It would be cheaper, more direct, and probably more effective if the government just ran advertising campaigns to "*Take shorter showers, disconnect your air conditioning (which accounts for over 30% of domestic energy consumption and 13% of commercial usage), hang your clothes out to dry instead of using your electric dryer, don't overheat your home in winter, and turn off your lights when you're not in the room.*" But most politicians just dare not use those words.

Some propose population control as a solution to global warming. The idea of controlling demand instead of increasing supply is appealing because it is less expensive and potentially less challenging. But in fact, it solves little and amounts to a feeble kicking of the electrical bucket a very short distance up the road. What is needed is a new source of environmentally friendly power.

Failure of fossil fuels to meet the planet's minimum energy needs

Since the beginning of recorded history, the most basic needs of a people have always included food, water, and shelter. In 2021, electricity is vital to all three.

Africa suffers by far the worst conditions of energy poverty. Eight African countries (South Sudan, Chad, Burundi, Malawi, Liberia, Central African Republic, Burkina Faso, and Sierra Leone) spanning over 3 million sq. km (over 1.2 million sq. miles) with a total population of over 100 million people are only able to supply electricity to between 5% and 15% of their citizens. In the other African countries, another 500 million people are without access to electricity.

Lack of access to modern energy services severely impacts the lives of a population, denying them basic needs such as heating, cooking, and lighting as well as drinking water, refrigeration, communication, education, and health services. Africa does not have a monopoly on such shortages. In the European Union, several million people in Spain live in a condition of energy poverty. Across the world, 1 billion people lack access to reliable electricity. And they all want it.

Food scarcity is the result of a vicious cycle beginning with water. Food production is directly dependent on water availability. Agriculture is water intensive; global warming is creating increasing areas of arid land that lack the water resources necessary to support the growth of sufficient produce and livestock. So, people starve.

Across the world, one in nine persons face hunger every day for themselves and their families. Africa is facing its worst food crisis in 75 years, with drought the key cause. Tonight, over 250 million people in Africa will go to bed hungry.

In the western hemisphere, the highest level of hunger is found in Haiti, where 25% of the population cannot be sure where their next meal will come from. Just so it is clear that hunger is not limited to Africa, note that over 10% of U.S. citizens (including 18% of the nation's children) also face hunger. The U.S. Supplemental Nutritional Assistance Program (SNAP) spent $60 billion in 2019 to provide free food for over 38 million Americans. When Covid-19 struck, food banks across the country saw a 40–50% increase in desperate people seeking food for their families. The world's wealthiest economy is not able to provide enough affordable food for its own people. And one way or the other, it traces back to the lack of a cheap, clean, and efficient energy source.

Despite being literally surrounded by the waters of the seas and oceans, drinking water is unavailable to a billion people on Earth **only because the technology that is able to process potable water from seawater requires more energy than is available**. Electricity generation from fossil fuels has shown itself to be incapable of meeting this need. And the wind and solar industries have thus far been unable to take up the slack.

To summarize, we have been married to a 19th-century energy source (fossil fuels) that has been a major contributor to the warming of the planet. At the same time, its emissions are poisoning the planet. And to top it all, its peak capacity is way below that necessary to provide the needed power, water, and food without which the Earth is kept under stress and unstable.

The cost of energy makes it harder to survive

Energy use is personal. Many of our day-to-day decisions depend in some way on energy and the price of energy—how we travel, what we eat, what temperature we keep our houses, and which jobs we work. Affordable energy makes our lives better because it facilitates all other economic endeavors.

For the past 25 years, the U.S. has spent an average of 1.25 trillion dollars a year on energy.[18] That is between 11% and 15% a year of the GDP. In less industrialized countries, that percentage is a little less, but in all cases climbing rapidly. U.S. residential users are charged twice the price for their electricity than are industrial users. Paying the electricity bills has become a problem for many families. Around 15% of senior citizens went hungry in order to be able to keep the electricity on. Electricity costs affect every sector of the economy.

Affordable energy would help to heal an ailing economy because affordable energy facilitates economic growth. Energy's share of GDP is one measure of the relative importance of energy in the overall economy. A new source of clean renewable energy that is affordable would allow heating, manufacturing, and transportation costs to plummet. In fact, pretty much all costs would plummet.

Is BP the answer to Global Warming?

As the world continues to heat up, the world's top 50 oil and petroleum companies continue doing what they have always done: prospect and extract, refine, transport, and burn oil and natural gas. These companies mainly stay out of global warming discussions except when they are lobbying against the imposition of restrictions on the fossil fuel industry. Their current message is a stern warning that the current underinvestment in fossil fuels will result in energy supply shortages. British Petroleum, LLC is the exception. One of the world's largest oil and natural gas companies with assets of over $282 trillion,[19] BP is trying to reinvent itself.

No longer is BP the perpetrator of the Deepwater Horizon oil spill (also known as the BP Oil Spill) in the Gulf of Mexico in 2010. No longer is it the same BP that pleaded guilty to 11 counts of manslaughter, two misdemeanors, and a felony count of lying to the Congress to cover up its complicity in that debacle, by far the

[18]EIA

[19]S & P Global, Platts Rank 2019.

largest marine oil spill in the history of the petroleum industry. Now, 9 years later, BP's CEO says, "It's time for us to tell our story a little bit differently."

And good as his word, over the past two years, BP has spent $50 million on its first global advertising campaign since the BP Oil Spill. The campaign showcases the company's efforts to embrace clean energy by placing advertisements on billboards, newspapers, and on major TV networks in the U.S., U.K., and Germany disseminating its message that BP is working "to make all forms of energy cleaner and better." The company went even further in its June 2020 Report where it announced: *"At BP, we are committed to playing our part (in reducing global carbon emissions). In February, we adopted a new purpose – to reimagine energy for people and our planet. And we announced a new ambition, to be a net (carbon) zero company by 2050 or sooner and to help the world get to net (carbon) zero."*

One of the world's largest petroleum/gas companies to have zero-carbon emissions? Wow!!

BP would be a welcome member of the alternative energy team if they had put their money somewhere other than where their oil wells were. In 2018, in just six oil/gas projects we could find, BP invested over $13 billion.[20] In 2019, in just five oil/gas projects we could find, BP invested over $20 billion.[21] BP's CEO bragged that over those 2 years, the company had invested $500 million "in low carbon technologies." We should not totally demean such an investment since $500 million is more than the U.S. government invested on alternative energies in the 4 years between the Obama and Biden administrations. But that $500 million is only 1% of the investment BP made in only 11 of its new oil well projects. This is not evidence of a company that is part of the global warming solution and racing to renewables.

What kind of future is BP building? More than 96% of the company's annual capital expenditure is on oil and gas. It is investing 30-1 in favor of high carbon emissions from fossil fuels. With respect, that is not going to get BP anywhere near to becoming a net zero-carbon emissions company, much less helping the world confront

---

[20]Clair Ridge (North Sea); Western Flank B, Thunder Horse Northwest Expansion, Shah Denis Stage Two, Taas-Yuryakh Expansion, Atoll Phase One

[21]Aligin (North Sea), Angelin (Trinidad), Constellation (U.S. Gulf of Mexico), Culzean (North Sea), West Nile Delta—Giza/Fayoum.

global warming. If BP really wants to get on board the 21st-century energy revolution, it needs to immediately start phasing out new oil and gas projects and increase its investment in alternative energies from $500 million per year to $500 billion. Its technology division could immediately start working on a real Energy Miracle.

The Seas Are Rising. Do Something!

You do not need three PhDs in science to notice something may be wrong with our climate. You can observe that the summers are getting hotter, tropical storms are getting more frequent, and more "super storms" are happening year by year. Figure 34 (top to bottom) shows increase in North Atlantic tropical storms, rising ocean temperature, and annual global sea levels rise.

Sea levels are rising. Tides are inching higher. High-tide floods are becoming more frequent and reaching farther inland. If you live by the seaside, you can observe the tides intruding further and further on shore. While all coastal cities will be affected by sea level rises, some will be hit much harder than others. Asian cities will be particularly badly affected. About four out of every five people impacted by sea-level rise by 2050 will live in East or Southeast Asia. U.S. cities, especially those on the East and Gulf coasts, are similarly vulnerable. More than 90 U.S. coastal cities are already experiencing flooding, a number that is expected to double by 2030. Meanwhile, about three-quarters of all European cities will be affected by rising sea levels, especially those in The Netherlands, Spain, and Italy.

Some of the cities most vulnerable to flooding in the U.S. include:

- **New Orleans, Louisiana** is already sinking. Hurricane Katrina killed more than 1600 people in 2005, leaving 80% of the city underwater.
- **Miami, Florida's** sea levels are already rising fast enough to damage homes and roads. A 2018 report from the Union of Concerned Scientists suggested that 12,000 homes in Miami Beach are in danger of chronic flooding within the next 25 years. That puts around $6.4 billion worth of property at risk. Other projections show that a 2-degree global temperature rise would submerge the entire bottom third of Florida—the area south of Lake Okeechobee—home to 7 million people.

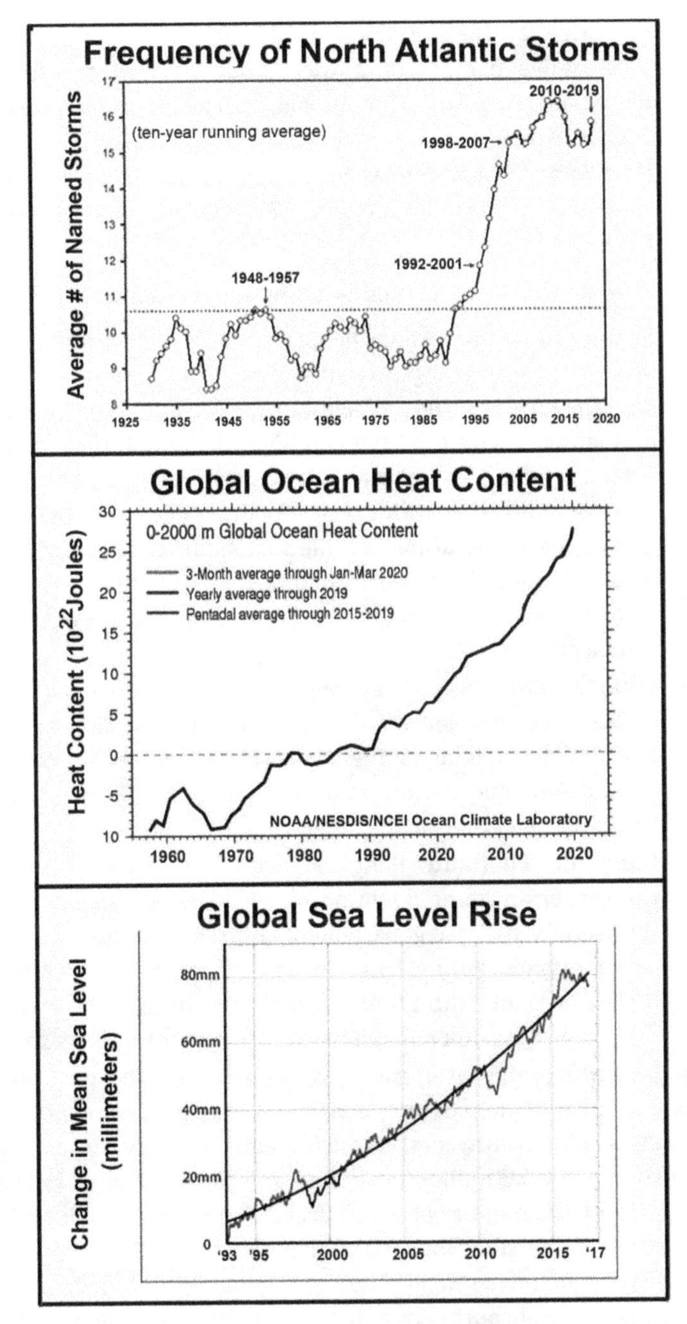

Figure 34 Storms/ocean heat/sea-level rise comparison (compiled from NOAA data).

- **Atlantic City, New Jersey** is no stranger to coastal flooding. Seventy-five percent of it was underwater as a result of Hurricane Sandy in 2012. In some areas, the water was as deep as 8 feet.
- **New York City** has long worried about the effects of rising sea levels. According to projections, 37% of lower Manhattan will be at risk from storm surges in the next 25 years.

Other cities seriously threatened by rising sea levels include:

- **Shanghai, China:** Besides being China's financial capital, it is one of the world's biggest ports and its most vulnerable coastal city. The former fishing village, bordered by the Yangtze River on the north with the Huangpu River dividing it through the center, includes several islands, two long coastlines, and miles of canals, rivers, and waterways. In the event of another 2–3 degrees of temperature rise, the vast majority of the city will be submerged in water, including much of the downtown area and both airports. Around 17.5 million people would be displaced.
- **Jakarta, Indonesia:** At a rate of 25.4 cm (approximately 10 inches) per year, Jakarta is the world's fastest sinking city. Much of this sinking is due to the digging of wells to access groundwater and also due to the weight of its buildings. As sea levels rise, Jakarta is increasingly under threat.
- **Bangkok, Thailand:** Bangkok faces a similar problem of skyscrapers pushing down on water-depleted soils. A study released by the city government in 2015 predicted it could be underwater within 15 years. The government is pumping water back into the ground, which has slowed the rate of sinking, but it is not enough to save the city from rising seas.
- **Singapore:** Most of Singapore is less than 15 m above sea level, and 25% of the country, consisting of land reclaimed from the sea in recent years, is considerably lower. Every year this country of 5.8 million experiences flash floods after intense rainfall, the number of which is ever increasing. A $75 billion fund has been established to maintain key infrastructure at least 3 m above sea level, with the most critical components such as the airport kept at 5 m above sea level. The country's leaders are all too aware that the "flooding models" cannot be

counted on, and there is no guarantee their actions will stay ahead of the rising seas.

- **Osaka, Japan:** Osaka is already confronting the threat posed by global warming floods. If temperatures rise another 2–3 degrees, Osaka would disappear beneath the waves, threatening some 6 million people.
- **Alexandria, Egypt:** Alexandria could be largely under water with an even lower temperature rise, affecting 3 million people in that historic city.
- **Rio de Janeiro, Brazil:** This city will have large sections flooded when global temperatures rise just a few degrees. The flooding will affect 1.8 million people.
- **Kolkata and Mumbai, India:** These are two cities severely threatened by sea-level rise. Other endangered Indian coastal cities are Surat and Chennai. In 25 years, it is predicted that 40 million Indians will be affected by flooding.

Rising ocean temperatures are the main cause of rising sea levels. Besides the cases mentioned earlier, huge urban populations are at risk in Guangzhou, China, Vietnam, Myanmar, Thailand, and Bangladesh; also, in many European countries. Dozens of islands and some entire countries could just disappear.

We ignore these warming signs at our peril. The conservative scientific consensus is that a 1.5°C increase in global temperature will generate a global sea-level rise of between 1.7 feet and 3.2 feet. Even if we manage to keep global temperatures from rising above 2°C, by 2050 at least 570 cities will be exposed to rising seas and storm surges. At risk are some 800 million people, their real estate, and their infrastructure, including roads, railways, ports, underwater Internet cables, farmland, sanitation and drinking water pipelines and reservoirs, and even mass transit systems. Some coastal cities and nations will literally disappear. The rest will need to adapt, and quickly.

The World Economic Forum's *Global Risk Report 2019*[22] shows around 90% of all coastal areas will be affected to varying degrees. Some cities will experience sea level rises as high as 30% above the global mean. Making matters worse, large cities are sinking under the

---

[22]World Economic Forum. *The Global Risks Report 2019* published by WEF Global Risks Initiative. Can be downloaded from the WEF website.

weight of new construction combined with the reduction in ground strength resulting from large amounts of groundwater extracted by their residents. (In parts of Jakarta, a city of 10 million people, the ground has sunk 2.5 m in the past decade. Chinese cities have been similarly impacted.)

The seas are rising. This is the major consequence of global warming and cannot be ignored. So what do we do about it?

Chapter 4

Roadblocks to an Energy Revolution

Existing Global Energy Sources

What follows is a list of the world's top 18 energy sources, listed here in the order of their current contributions to the world's energy production and their future prospects.[23]

1. Oil
2. Coal
3. Natural gas
4. Hydropower
5. Nuclear
6. Biomass
7. Wind power
8. Solar photovoltaics (PV)
9. Active (concentrating) solar thermal
10. Passive solar
11. Geothermal energy
12. Energy from waste
13. Ethanol

[23]Richard Heinberg, *Searching for A Miracle: Net Energy Limits and the Fate of Industrial Society,* International Forum on Globalization and the Post Carbon Institute, False Solution Series #4, Sept. 2009.

Energy Miracles: The Global Warming Backup Plan
H. B. Glushakow

ISBN 978-981-4968-18-8 (Hardcover), 978-1-003-28442-0 (eBook)
www.jennystanford.com

14. Biodiesel
15. Tar sands
16. Oil shale
17. Tidal power
18. Wave energy

Coal, crude oil, natural gas, and combustible renewables account for 91% of the total. Nuclear constitutes an additional 5%, and hydroelectric, another 2%; while all the others (including wind and solar) comprise 2% of current energy production. Richard Heinberg, after 30 years of studying them, gives a rather grim assessment of the ability of any of these to meet growing global energy needs. Even grimmer is his analysis that none in the bottom 14 will be able to replace any of the top three fossil fuel sources any time soon. He chalks that up to the relatively cheap cost and relatively high energy density of the top three (oil, coal, and natural gas), meaning fossil fuels are much more efficient methods of generating power. But that is not saying much. When these fossil fuels are running steam turbines, they lose 65% of their prime energy as heat right out of the gate.[24] That does not take into account the energy consumed in constructing and running these mammoth power plants, or in locating, extracting, and refining the fuels used. We can do much better than that. The real reason we have been stuck with these technologies is that scientists have been looking for an answer by peering down the wrong coalmines. We need to dig deeper into our energy bullpens to find real 21st-century Energy Miracles.

Is the Answer to Economize?

From the data given earlier, it is clear that despite three decades of marketing and promotion plus massive investments and price guarantees from governments, the world is not generating enough power from renewable energy sources to meet anywhere near the most minimal demands. That is true for every single country.

Against that background, we might still reduce fossil fuel emissions by applying an approach usually considered logical when

[24]Electropedia: Energy Efficiency. https://www.mpoweruk.com/energy_efficiency.htm

dealing with situations of scarcity. You have only $10 left in your pocket, and you will be getting your next paycheck in 3 days. For dinner tonight, you and your partner will share a couple of leftover dishes and limit your beer consumption to one each. This is called "economizing," and it is always possible to find ways of reducing personal and corporate energy consumption.

Ask any designer of off-grid renewable energy systems and he will tell you that the first thing to confront when contemplating the switch from the local electricity company to renewables is the need to economize on energy usage. When someone decides to move off the grid and run a household exclusively on wind and solar power, his first decisions will be which electrical appliances he will be able to discard and how he will limit the amount of energy being used by those he cannot do without. Things like: get used to working by the light of one bulb when you used to keep five burning; no central air conditioning; heat the room you are in, not the entire house; use one-tenth the amount of hot water you are used to by bathing in lukewarm water and limiting showers to a minute or two the way it is done in the Navy; use one TV instead of three and make it energy efficient with a small screen; run one car instead of two and use car pools and bicycles whenever possible.

Those solutions do not sound very sexy and many in the affluent West may not be willing to go along with them. Most Americans would consider it a violation of their human rights to be restricted to one car or a 2-min shower. Remember, the main Trump disagreement with the Paris Accord was that it "punished Americans with onerous energy restrictions." The middle classes in the U.K., France, Germany, and China largely agree. Politicians rarely dare to bring the subject up.

In an earlier time, this was called conservation of resources or "being economical." Nowadays, it is largely missing from the "renewable energy" conversation. It is as though our leaders believe it is perfectly ok for people to waste as much energy as possible (while expecting the government to keep subsidizing energy prices). You see people running around their houses in winter dressed in their underwear because the heating system is overheating the house, and the same people wrapping blankets around themselves during the summer, because the air conditioning has turned the house frigid.

Energy efficiency has been a big buzzword in industry for the last couple of decades. Products that save energy should have been big hits. Variable speed drives (VSDs) or variable frequency drives (VFDs) are an example. Forty percent of electrical energy is used by industry, and two-thirds of that is used by electric motors. VSDs, which regulate the speed of a motor and allow the motor to run more smoothly, can reduce the motor's energy consumption by up to 50% in many cases. Yet less than 20% of motors are equipped with VSDs.

As a society, we are often oblivious to waste in its many forms. It is up to civic and community leaders to call this subject to attention, and up to every citizen to enforce conservation in his immediate environment. But, putting aside the issue of waste, this is one of those times that it is not possible to economize ourselves into affluence. No matter how much we may cut back on energy usage, this action will not solve the world's energy problems; neither will it solve global warming. That is because too much energy is needed and that problem is not going away. Thirty years from now, the world will be consuming at least 60% more energy than it does today.

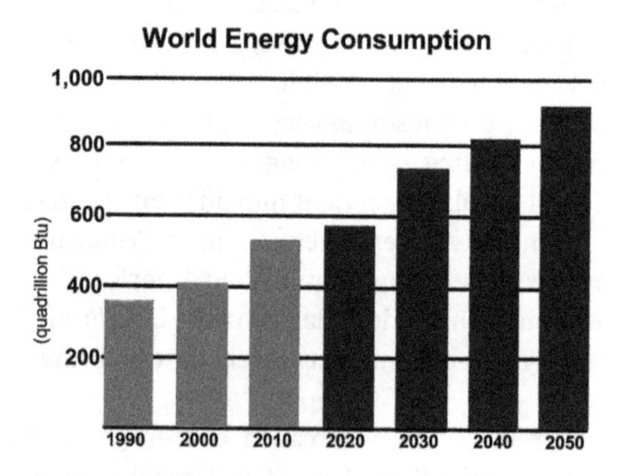

Figure 35 World energy consumption in 2050 (data extracted from Statistica).

2020 Global Renewable Outlook and Initiatives

The International Renewable Energy Agency (IRENA) released its *Global Renewable Outlook 2020* in the midst of the Covid-19 pandemic. It argues that the way forward to 100% renewable energy will require that government "stimulus and recovery packages accelerate the shift to sustainable, decarbonized, economized economies." The same report states that up to $1 trillion of new investment would be required to meet its goals, yet admits that not only is that not happening, but renewable energy subsidies are actually falling.

IRENA has a range of plans to reduce carbon emissions to zero (some quick, and others less so). These plans promote the widespread adoption and sustainable use of all forms of renewable energy, including bioenergy, geothermal, hydropower, ocean, solar, and wind energy. But the three energy sources on which IRENA's plans most heavily rely are wind, solar, and hydropower. The problems of wind and solar were enumerated earlier. As for hydro, a report from the Environmental Defense Fund[25] (EDF) points out some little-known facts about hydro that can cause projects meant to reduce greenhouse gas emissions to unintentionally increase them instead. Its 2019 study of 1500 hydropower plants from around the world found some hydropower plants are having a greater warming effect on the environment than fossil fuels. Not in every case, to be sure, but mentioned here to illustrate the fact that even hydropower may not be as carbon free as we have usually thought. Not only that, but most of the prime sites for hydroelectric plants in Europe, the U.S., and China have already been taken. You cannot build another dam across the Colorado River in the U.S. or at the Three Gorges in China. IRENA and EDF and Al Gore and Bill Gates (to name just a few) all agree with the statement, "While each country must work with a different resource mix, all of them need a 21st-century energy system."

Redundancy: 21st Century Energy Innovation to Back up Wind and Solar

Redundancy is a word that can be a total turnoff because its common usage implies either someone who is no longer useful and about to be fired or a part of a system that has no use or function. Redundant actually comes from two Latin words meaning *a wave that rises again.*

---

[25] I. B. Ocko and S. P. Hamburg, Climate impacts of hydropower: enormous differences among facilities and over time, *Environmental Science & Technololgy*, **53**, 23, 2019, 14070–14082.

That is also its connotation when used in engineering: something designed into a system that can step in and operate in case another part of the system fails. It is not something useless. Rather it is a backup, something that can restore a system to functionality and instantly bring it back to full strength after a breakdown.

Almost every security or safety system employs redundancy, as does every modern bridge, every airplane system, elevator, and maritime navigation system. If a pump failure would destroy your entire factory, it is prudent to have two pumps, the second set to automatically start up in the event of malfunction or damage to the first. If a critical data system would crash without a power supply, built-in redundancy would include a second power supply. Cars with hydraulic braking systems will also have a mechanical braking system in case the hydraulics fail.

And so, we come to the search for alternative energy sources. The world has pretty much accepted a basket of three alternative energy sources (wind/solar/hydro) to replace the fossil fuels, though it is acknowledged that there are problems with deploying any of these fast enough and broadly enough to meet the current challenge. With the consequences of global warming being so severe, can we afford to place all our hopes for the future in one basket?

Figure 36 Can we afford to place all our eggs in one basket?

Abundant electricity is needed to support our industries, cultures, and lifestyles. For 150 years, we have been getting that electricity from giant gas and coal-burning power plants that are warming up the planet at the same time they are poisoning it. In addition, the consequences of global warming plus the growth of the world's populations require much more power to be generated than those fossil fuel generators can possibly produce.

In 2019, the world's coal consumption fell, yet was still the single largest source of power generation, accounting for over 36% of global electrical power. In 2018–19, the average annual growth in carbon emissions was greater than its 10-year average.

This is where redundancy comes into play—to back up the existing alternative energies. While doing everything possible to assist, expand, and forward wind, solar, and hydropower capacities, we need a second plan, a backup plan, a Plan B, to ensure that the challenges of global warming will be met, **no matter what**.

If the consequences of global warming are as dire as is believed, can we afford to depend only on wind, solar, and hydro to face the global warming challenge? What is needed is a 21st-century energy revolution.

Why are we so addicted to 19th-century power sources? What blocks are standing in the way of real innovation?

That is where we are going in this book. It has to do with the subject of energy.

Chapter 5

The Energy Epoch Starts Now

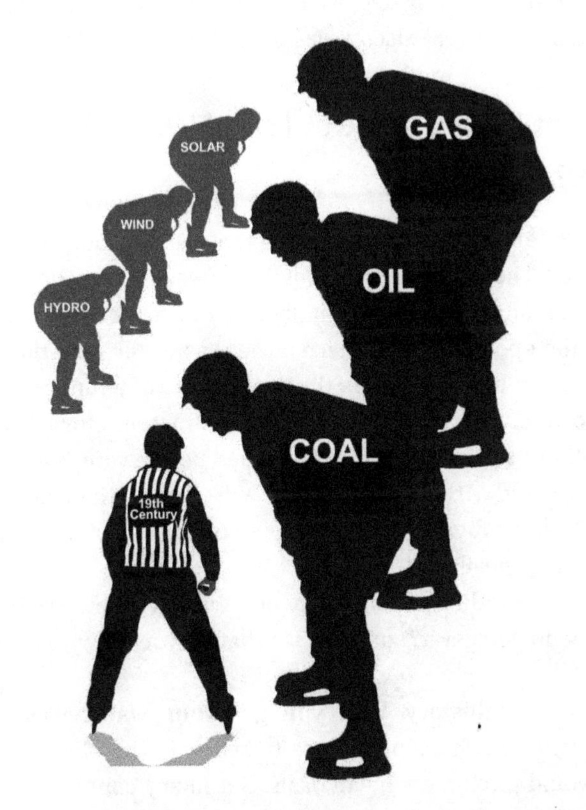

Figure 37 The 19th-century energy sources face-off.

Energy Miracles: The Global Warming Backup Plan
H. B. Glushakow

ISBN 978-981-4968-18-8 (Hardcover), 978-1-003-28442-0 (eBook)
www.jennystanford.com

We are strongly routing for the guys on the left. But the outcome is not inevitable. And the main question of this book is why Earth requires that every participant in the 21st-century power sources game be a relic of the 19th century?

There is a subject called "electricity" and a profession called "electrical engineering," which should have provided innovative solutions to this problem decades ago. Why have not new energy sources been found or developed or invented?

This chapter begins to explain, starting with the subject of electricity, then moving to electrical engineers, and finally surveying the greatest electrical discoveries in history.

Electricity: How You Get It and What You Do with It Once Got

We know electricity as the stuff that comes out of electrical generators. There are useful analogies between water and electricity, but not the ones that you normally learn in school. Electricity is as vital to the operation of modern society as water is to the survival of the human body. Whereas there are cases of people surviving a month or more without food, water is a different story. Just 3 or 4 days without water are all that is needed before your body starts to shut down. And without electricity, it will take less than half a second for trains and subways to fail, lights to extinguish, and TVs and the Internet to go eerily quiet.

There is another analogy between water and electricity. There are "How do you get it?" and then "What do you do with it once you have it?"

Throughout history, **how you got your water** was a major consideration in choosing a place to settle. In the Middle East, the smart would pitch tents by an oasis. But in any country, in any age, before building a house you would factor in whether the land had a lake or pond or natural spring water; whether it was near a river or stream; or whether the water table was such that a well could be dug to bring water up from below the ground.

Once you had the water, then the second factor went into play: **what you did with it**. Most rudimentarily, your daily chores would include carrying buckets of water up from the stream to the campfire for cooking. Water would be necessary to grow your crops and keep your animals alive. Water could be used to transport your harvested products down to market. You could make tea or coffee with it and wash in it. Good water is a vital ingredient in the process of distilling beers and whiskeys. Eventually, you could start channeling water directly to your house by gravity or by the use of pumps. Getting fancier, you could plumb a house, so the water went directly to your cooking area, toilet room, and then multiple outlets, hot and cold. Outside, hose outlets allow you to water your crops or lawns. Irrigation systems can automatically adjust for amount of water dispensed and type of spray and be set to turn on and off according to a defined schedule or how dry the soil becomes. Water density in your home's atmosphere can be adjusted whenever it becomes too damp or too dry.

All of these things can now be operated from a smartphone. These can be complex processes, but all fall into either one or the other of these two categories: (1) how you get it and (2) what you do with it once got. The same two steps hold true for electricity.

The basic thing you get is called "electricity." What you do with it once got is called "electronics." These days more than 99% of electrical and electronic engineers are involved in the "what you do with it once gotten" sector. They are busy designing, developing, testing, and manufacturing new or advanced equipment for coal or gas-burning power plants, electric motors, broadcast and telecommunications systems, navigation systems, smartphones, computers, automated equipment, etc. Far less than 1% of them ever think about new ways to produce the electricity being used. How vibrant is the electrical engineering industry? According to the U.S. Department of Labor Statistics, even before the Covid-19 pandemic, the occupation of electrical and electronic engineers was already stagnant with zero growth expected.

Natural electricity is familiar to everyone as the giant flashes in the sky we know as lightning. At any given moment, there are 2000 lightning storms active around the world producing 100 lightning strikes per second. All lightning is direct current traveling in a single direction. In the 18th century, Benjamin Franklin discovered that a high potential energy (electric charge) could build up in the clouds looking for a place to go. At the ground level of Earth, there is almost a total lack of this potential (zero electric charge), which creates a vacuum beckoning to the charge in the clouds to come fill it. Franklin called the two sides of this dichotomy positive and negative charge. When the potential in the clouds becomes great enough, it seeks out its opposite on the ground and kaboom! Lightning flashes.

Figure 38 Lightning occurs when the zero potential on Earth attracts the high-energy potential in the cloud.

Other examples of electricity in nature include static electricity and the electric eel. Both are also direct current flows (DC) that work on the same principle as lightning. The first commercially produced electricity was copied from this pattern. Artificially created electricity is a relatively recent occurrence as things go on Earth. In 1831, scientist Michael Faraday discovered that when a magnet was moved in relation to a coil of wire, an electric current flowed in the wire. Forty years later, James Clerk Maxwell discovered that magnetism and electricity were just different statements of the same phenomenon, both created by the same force. He found that not only moving a magnet in relation to a copper wire would cause electrical current to flow through that wire, but an electric current moving through a wire would also produce a magnetic field.

Amazingly, that is pretty much all there is to power generation, and it has been so for 150 years. To keep all those trains and refrigerators running, all those smartphones and iPads charged, and the Internet and airlines online and functioning, all that we have needed to do was find ways to get a copper wire or an iron core rotating inside a magnet. It works just as well if you rotate magnets inside copper coils in the generator, which is what most large power plants do. Since the 19th century, there have been only four basic components to the mechanism:

1. The magnet
2. The wire that rotates
3. The mechanism that causes the wire to rotate
4. Essential, but neglected in textbooks, is the base that keeps the poles apart[26]

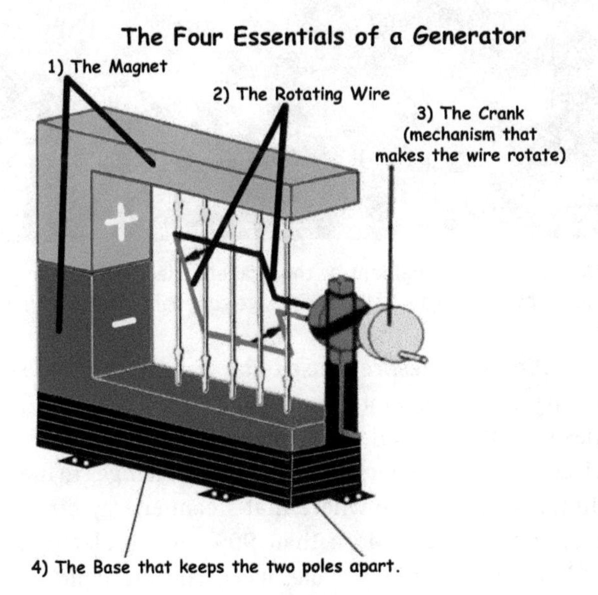

Figure 39 The four essential parts of a generator.

[26]More on this point in later chapters.

Turning the Crank: Planet Earth's Obsession with 19th-Century Power Generation

The essence of 90% of global electrical production today and the problems and limitations associated with it lies in the third component in Fig. 39: the crank. The phenomenon of electricity may be wonderful, but where it relies on a mechanical device consisting of one part rotating within another, it is not going to work without a crank. In other words, you need a way to make those wires rotate within the magnetic fields. Hand cranks would not cut it. In large generators such as the one that can be seen in Fig. 40 from GE Steam Power, you are talking about rotating parts that weigh over 1000 tons.

Figure 40 The rotating element of the Arabelle electrical steam generator, manufactured by GE, weighs 1215 tons. Image copyright by GE Steam Power.

The solution in widespread use today has not changed since the 19th-century steam locomotive.

In these beautiful Iron Horses, coal heats up water, which creates pressurized steam. The energy thus produced changes to mechanical power in the steam engine where that steam energy directly turns the wheels of the train. More than 90% of the electrical power generated in the world today makes use of this principle.

1. Water is heated up by burning oil, coal, natural gas, or "bio" materials.

2. The steam pressure produced "turns the crank" shown in Fig. 39. In power generation plants, the cranks are called turbines.
3. "Nuclear Power" is exotic, and by far the most efficient method of heating the water, but it works on exactly the same principle.

Figure 41 Steam locomotive (19th century).

Hydroelectric power does not require heat or steam since the force of a flowing river directly turns the crank. The same for wind power generators where the wind provides the force to turn the crank.

Solar (photovoltaic) power is different. Photovoltaic uses a chemical reaction to directly make current flow from the heat of the sun. Traditional solar cells are made from silicon, the result of putting sand through a complex purification process. Silicon is used because it absorbs a wide range of the sun's light, conducts electricity well, and can be easily processed into very thin silicon wafers, the heart of a photovoltaic cell. Sunlight shining on the solar cell separates into plus and minus charges. This starts a flow of electrons, which is electricity.

But why in the 21st century are we still relying on 19th-century power generation methods?

I know you can go on YouTube and find out how the big petrochemical companies have been buying out or intimidating the inventers of alternative energies for their own selfish profits. I do not doubt some of that has gone on, but blaming our current situation on the ghost of John D. Rockefeller, Sr. or on Exxon Mobil does not lead to a resolution of our problems.

Do we really need to continue our reliance on the 19th-century crank?

The reasons we are still tied to that crank and the path to breaking those ties are revealed in the next few chapters. But first we will review the role of the electrical engineer and survey the greatest electrical discoveries of all time.

Electrical engineers: they lost the E in EE

EE is the abbreviation for electrical engineer. This has been a worthy discipline since 1884 when some of the most prominent members of the then fledgling field of electrical engineering (including Nikola Tesla and Thomas Edison) formed the American Institute of Electrical Engineers (AIEE) with the purpose *"to promote the Arts and Sciences connected with the production and utilization of electricity."* In 1891, Alexander Graham Bell stepped down as president of the AIEE and chaired a committee to adopt a logo for the organization.

Figure 42 American Institute of Electrical Engineers logo.

To make the mission of the electrical engineer perfectly clear, the logo chosen shows a magnetic compass, Ohm's law and Benjamin Franklin's kite, memorializing the major discoveries at the core of electrical knowledge and the procedures by which they came to be discovered, including Franklin's experiment proving that lightning was electricity. (The formula below the compass is Ohm's law in the notation of the time.)

The end of the 19th century, when the AIEE was formed in New York City, was a supernova of discovery and innovation in the subject of electricity. In the quarter century following the 1860 release of Maxwell's famous laws unifying magnetism and electricity, power generation went from theory to actuality with the construction first of DC and then AC power production plants, and hydroelectric power plants going on line first in Britain and then in the U.S. New York City got the first broad scale rollout of electrical power for residential and industrial use, but it quickly expanded to the rest of the U.S.

Electricity and electrification did not just "happen." It developed as the result of a fury of research, innovation, and debate. The Direct Current guys versus the Alternating Current guys were in a mad race to prove their concepts and get their products to market. It was called the War of the Currents, and industry was investing big money into the race. Universities were researching it, and new science graduates were being gobbled up by the many corporations seeking to extend its development. It was a noisy and colorful time. On one side, Thomas Edison was broadly promoting his direct current electrical system. On the other, George Westinghouse and Nikola Tesla (a brilliant inventor who had been fired by Edison when he pointed out the shortcomings of using DC current) were promoting their alternating current system. Edison tried his colorful best to discredit alternating current technology by staging public events where dogs and horses were killed using Tesla's alternating current machines. (Edison bought stray dogs from neighborhood boys in Orange, New Jersey at 25 cents apiece for the purpose.) Edison graphically demonstrated how AC "electric chairs" could kill a man convicted of capital crimes "in the ten-thousandth part of a second" by arranging for a criminal at New York's Auburn Prison, William Kemmler, to be executed by the first electric chair (which Edison ensured was being powered by one of Tesla's AC machines). In 1903 on a stage in Coney Island, New York, electricians from the Edison Company killed a live elephant on stage with "Westinghouse AC power." The Edison Film Company filmed it with the movie's screen credit going to: "Thomas A. Edison." Its message was clear: "Is this the type of electricity your wife should be cooking with?"

Figure 43 Edison's filmed electrocution of Topsy the Elephant, New York, 1903.

Despite all of Edison's showmanship, he ended up failing when it became clear that alternating current solved many of the limitations of Edison's direct current machines. But AC systems were more complex, and the rapid growth of the power industry in the 1880s and 1890s generated an unprecedented demand for trained electrical engineers. University physics departments met the demand by adding staff and facilities to handle the influx of students.[27]

IEEE

Sometime in the late 1930s, the subject of electrical engineering lost its way. Less than 25 years later, in 1963, the American Institute of Electrical Engineers added an E to its name, making it the Institute of Electrical and **Electronic** Engineers (IEEE). Since then, the focus of the organization has shifted steadily from the electric to the electronic.

Spectrum, the monthly journal of the IEEE, is beautifully laid out and illustrated, but has shown little to no interest in the search for new sources of energy or electricity. In a survey of almost

[27]Robert Rosenberg, American physics and the origins of electrical engineering, *Physics Today*, **36**, 1983, 48–54.

400 articles appearing in its pages over the last 4 years, only four concerned research into new energy sources. There were many more articles from psychiatrists, neurosurgeons, and accountants than from electrical engineers writing about electricity. The 178 articles that appeared in its last 18 months of issues featured subjects like Flying Car Sales, Medical Implants, Electronic Toys, Quantum Computing, Drones Delivering Blood, AI, Robots, Moon Habitats, and whether Animals can (or should) use the Internet. There was one article about Fusion Energy, but that was rather cancelled out by another outlining the public risk and problems of storing nuclear waste material produced as a by-product of nuclear power plants in Washington State. (So far, the rusty ageing nuclear waste containers have leaked 4 million liters—over 1 million gallons—of radioactive sludge into the Columbia River.)

In 18 months, there was just one article in search of better energy sources, written by retired NASA engineer Jay Schmuecker who bought his grandfather's farm and tried to turn it into a carbon-free operation. Of course, the farm needed a tractor and that tractor had to be retrofitted to run on his solar-hydrogen system. The system itself required generators to create pure hydrogen and pure nitrogen, a reactor to combine the two into ammonia, tanks to store all the gases plus 360 solar panels to run the system. That turned into an impossible task. He ended up spending $2 million building a system that could produce only 10% of the hydrogen and ammonia needed just to run the tractor. But the words with which he ended the article were inspiring:

> *"Humankind needs to develop renewable, carbon-emission-free systems like the one we've demonstrated. If we do not harness other energy sources to address climate change and replace fossil fuels, future farmers will find it harder and harder to feed everyone. Our warming world will become one in which famine is an everyday occurrence."*

The 2021 IEEE Vision, Innovation, and Challenges Summit and Honors Ceremony honored over 50 contributions to science over the previous 12 months. They included many brilliant and wonderful contributions to humankind and to technology, yet not a single one of them advanced the search for new sources of electrical energy.

A Lucky observation

The venerated Dr. Robert Lucky, fellow of the IEEE and member of the National Academy of Engineering, in a *Spectrum* article, remembered the soul searching that accompanied the incorporation of the word "electronic" into the IEEE back in 1963. Lucky voiced his concern over the subsequent evolution of the educational and professional institutions that support engineers and the perception of electrical engineering currently held by potential students.

He asked these searching questions:

What are those things that all EEs should know?

What is the commonality of training and experience that holds EEs together as a profession?

In his poignant words: "How will electrical engineers be distinguished from other branches of engineering?"

Putting some electricity back into electrical engineering

Dr. Lucky's questions are answered at the end of this chapter, but his observation was spot on. To get a degree in electrical engineering (EE) these days, you are ordinarily obliged to do so in a department named **Electrical** and **Computer** Engineering (ECE). The curricula in these departments of ECE are severely oriented toward the computer side. This is not meant as criticism—computer engineering is obviously essential to our culture—but slighting the electrical side has been a serious omission, and directly impacts our ability to find alternative energy sources.

Figure 44 Sign in front of University of Illinois, Urbana-Champaign ECE Building.

The term "electrical engineering" has not totally disappeared. There are still schools around that grant EE degrees. But most of them interpret EE the way that Trine University does.

Trine University, in Angola, Indiana, is one of many colleges and universities with real campuses that offer actual degrees in electrical engineering. The program offered is explained as follows:

As a Bachelor of Science in electrical engineering student at Trine University, you will learn about the application of electrical energy for practical use. The ***electrical engineering*** *program is broad-based, covering the main aspects of electrical engineering:*

- *Generating and delivering power*
- *Communications and control*
- *Digital design*
- *Instrumentation*

The many schools with similar curricula offering "electrical engineering" degrees are creating technicians capable of running today's power plants and measuring all sorts of interesting things. However, they are not making progress in electricity technology. It would be more productive for those schools to extend their curricula to also create electrical engineers with the interest and tools to discover new sources of power for the 21st century.

A survey of the curricula of America's top six engineering schools found similar situations existing in all of them.

The University of Illinois Urbana-Champaign

The University of Illinois Urbana-Champaign introduces its electrical engineering program with this statement: "*The electrical engineering core curriculum focuses on fundamental electrical engineering knowledge: circuits, systems, electromagnetics, semiconductor devices, computer engineering, and design. The rich set of ECE elective courses permits students to select from collections of courses in the seven areas of electrical and computer engineering: bioengineering, acoustics, and magnetic resonance engineering; circuits and signal processing; communication and control; computer engineering; electromagnetics, optics, and remote sensing; microelectronics and quantum electronics; power and energy systems....Electrical engineering is a multifaceted discipline that over the last century has produced an astounding progression of technological innovations that have shaped virtually every aspect of modern life. Electrical engineers need ... education in the engineering principles of analysis, synthesis, design, implementation, and testing of the devices and systems that provide the bedrock of modern energy, communication, sensing, computing, medical, security, and defense infrastructures."*

Nothing in that statement would focus a student's attention on the basics of energy or new ways to produce electricity.

Carnegie Mellon University

The Carnegie Mellon University (CMU) is a global research university ranked as one of the world's top engineering schools. Its electrical engineering program is found in its Department of Electrical and Computer Engineering. The five areas of study within the ECE department are: device sciences and nanofabrication, signals and systems, circuits, hardware systems, and software systems. Electrical engineering courses include an introduction to soldering, printed circuit board layout and fabrication, computer systems, semiconductor devices, microelectronic circuits, computer and network security, neural stimulation and sensing technology, and hardware arithmetic for machine learning.

But nothing about understanding energy or finding new ways to create the electricity the planet so desperately needs.

Harvard University

Harvard, America's oldest institution of higher learning, describes its electrical engineering program as spanning "*a broad range of topics, ranging from the physics of new materials and devices, the circuits and next-generation computing platforms made from these devices, and the algorithms that run on these platforms. The range of subtopics includes power systems, microelectronics, control systems, signal processing, telecommunications and computer systems. Students learn how to analyze design and build devices and systems for computation, communication and information transfer.*"

In Harvard's own words, "*Electrical Engineering at the Harvard School of Engineering studies systems that sense, analyze, and interact with the world.*" Loudly missing is a study of basic energy that would permit of new, clean, abundant sources of electricity and power to feed our hungry billions.

Massachusetts Institute of Technology

MIT is the world's number one engineering institution. It was MIT, in 1882, that established the world's first electrical engineering program in its Physics Department with a curriculum heavily oriented toward power engineering. At that time, power engineering was a crusading, pioneering part of electrical engineering that included such things as Michael Faraday's discovery of electromagnetic induction, the design and construction of new types of DC and AC power stations, generators and transformers. Today, MIT's Electrical Engineering curriculum is found in its Department of Electrical Engineering and Computer Science (EECS). Its EE program is introduced in terms of robotics, communication networks, medical technologies, and interconnected embedded systems. EE courses include circuits and electronics; signals, systems, and inference; nano electronics and computing systems; electromagnetics and applications; electromagnetic fields, forces, and motion; and cellular neurophysiology and computing.

Students will study plenty of computer and electronics and consequential electronic applications, but these days there is a paucity of actual **electrical** in MIT's EECS.

Stanford University

Stanford has a Department of Engineering that grants EE degrees. In the university's own words, "The EE program includes a balanced foundation in the physical sciences, mathematics and computing; core courses in electronics, information systems and digital systems; and develops specific skills in the analysis and design of systems." Students will spend much of their time on integrated circuits, biomedical applications and power electronics, but also offered are majors in Physical Technology and Science. The reason for offering the broadened course structure, according to Stanford, is "*Research in energy is motivated at the macro level by the rapid rise in worldwide demand for electricity and the threat of global climate change and on the micro level by the explosion in the number of mobile devices and sensors whose performance and lifetimes are limited by energy.*"

That reasoning is sound. What is missing is a structured curriculum to encourage interest and progress in new energy technologies that can create the quantity and quality of electricity needed to save our planet.

University of California-Berkeley

At UC Berkeley, electrical studies are done in the Department of Electrical Engineering and Computer Sciences (EECS). Although the Bachelor of Electrical Engineering and Computer Sciences will not get you very far down the road to finding better sources of electricity, the **Bachelor of Science in Energy Engineering*** is more on point. This program "*interweaves the fundamentals of classical and modern physics, chemistry, and mathematics with energy engineering applications. A great strength of the major is its flexibility. The firm base in physics and mathematics is augmented with a selection of engineering course options that prepare the student to tackle the complex energy-related problems faced by society.*" One special topic is "*New devices and energy sources.*" Although not specifically focused on new energy, with such an approach, there is hope.

A final fact in case you're wondering if any of the above is exaggeration. I was able to identify 700 current faculty members with EE degrees in the six aforementioned universities. Most have their own web pages and their areas of interest are clear for all to see including things like AI, computer engineering, data and communication systems, automation and robotics. Amongst these 700, I could find no interest or activity in the search for an energy miracle other than from a score of courageous folks at UC Berkeley (comprising 3% of the total).

The work of art that summarizes it all

This work of art (Fig. 45) at the University of Illinois ECE (Electrical and Computer Engineering) Building in Urbana, Illinois, puts the missing "E" into perspective. This stark sculpture by John Adduci makes a powerful statement of the second-class status to which the subject of electricity has sunk by the second-class position to which it has been relegated, outside the back door of UIUC's state-of-the-art facility.

---

*Energy Engineering, a recent discipline to emerge in universities, in most places focuses on wind and solar energy, power distribution, new types of battery cells, economics, and energy-related public policy. All these are important, but more important is to give students a better grasp of the subject of energy.

Figure 45 Diss-connections at the back door of UIUC ECE building.

The irony of this much larger than life metal form depicting three **electrical connections** must have gone over the heads of the ECE department. Did they ask the artist whether he was contemplating the degree to which "electrical engineering" was being demeaned when he named his work: "Diss-Connections?"[28]

International Conference on Environment and Electrical Engineering

The 18th International Conference on Environment and Electrical Engineering was held in June 2018. The event is one of Europe's largest and longest running professional and technical exhibitions. In the 18 years it has been running, it has provided a platform where designers and industry interact with manufacturers, energy utilities people, and university researchers to discuss all manner of issues relating to energy systems and the environment. Broad? Yes, but not broad enough.

Of the 60 Technical Areas scheduled for discussion at the conference, 10% pertained directly to existing alternative energy technologies (wind, solar, biomass, hydro). The remaining 90% were

[28]Diss. Definition: to treat with disrespect or contempt.

limited to the application, control, and measurement of those existing alternative energy technologies. Not a single category targeted the search for a new, better, or more functional energy source.

The conference is sponsored by the Institute of Electrical and Electronic Engineers (IEEE) EMC Society (EMCS), Industry Applications Society (IAS), and the Power and Energy Society (PES).

» » » » »

This section is, by no means, meant as any disparagement of the IEEE or any of the fine schools mentioned. I am a Senior Member of the IEEE myself, and one of my sons got a fabulous education at one of the aforementioned departments of electrical and computer engineering. He graduated early with honors and was immediately whisked away by one of the tech giants for a gig in the upper stratosphere of computer engineering. The point of this section is only that, at this time, the discipline of electrical engineering seems to have lost its way. And we need it back on the rails to squarely meet the challenges of global warming.

Greatest Discoveries in Electricity

In terms of its benefit to humankind, the generation of electrical energy exceeds the impact of the discovery of the wheel or the management of fire. The progress of human's discovery and use of electricity is marked by many important milestones. Because the production of electrical power in greater volume (but with fewer detrimental side effects) is central to solving so many of today's challenges, we look briefly at the progress of electricity discovery and development.

Table 2 is a collection of the world's most important electrical advancements, listed in date order. In the course of its history, our planet has seen some amazing electrical engineers. Because a Brit is likely to favor Englishmen and a Berliner will favor Germans, no two people will exactly agree on whose discoveries merit inclusion in this "best of the best" list. Nevertheless, we have given it a shot here.

Table 2 Greatest discoveries in electrical energy

	Date	Inventor	Discovery
1	600 BC	Thales of Miletus	Discovered static electricity by rubbing amber with furs
2	1600	William Gilbert	Coined the word "electricity" from the Greek "*electron*" (the Greek word for amber)
3	1705	Francis Hauksbee	Discovered the neon light
4	1720	Stephen Gray	First to distinguish things that would conduct electrical charge and things that would insulate it
5	1745	Pieter van Musschenbroek	Invented the Leyden Jar—a means of storing electric charge, which could later be discharged at will
6	1752	Benjamin Franklin	Discovered the two-pole nature of electricity and that lightning was electricity
7	1783	Charles Augustin de Coulomb	Showed the attractive force between two electrically charged particles was dependent on the charge and the distance between the particles
8	1786	Luigi Galvani	Discovered electrical current flowed between different metals
9	1796	Allesandro Volta	Invented an electric battery capable of creating a continuous flow of electric current
10	1820	Hans Christian Oersted	Discovered that an electric current would generate a magnetic field
11	1820	André-Marie Ampère	Called "The Newton of Electricity" by Maxwell, he observed that currents would repel or attract each other depending on their direction of flow. Developed the first mathematics predicting the relation between electricity and magnetism and posited the existence of the electron as the basic component of both subjects

(*Continued*)

Table 2 (*Continued*)

	Date	Inventor	Discovery
12	1821	Thomas Johann Seebeck	Discovered that voltage (and current) flow between two dissimilar metals in proportion to their temperature differences
13	1821	Michael Faraday	Built the first electric generator based on his discoveries of electromagnetic rotation and induction. Further he found that electric induction takes place through an intervening vibrating medium (aether) and not by some incomprehensible "action at a distance"
14	1823	William Sturgeon	Produced an electromagnet
15	1826	George Ohm	Precisely defined the relationship between power, voltage, current, and resistance
16	1832	Hippolyte Pixii (and André-Marie Ampère)	Built the first direct current commutator-type continuous current generator
17	1833	Samuel Hunter Christie	Invented the Wheatstone Bridge for measuring electrical resistance
18	1834	Thomas Davenport	Invented the first battery-powered electric motor
19	1835	Joseph Henry	Discovered the electromagnetic phenomenon of self-inductance and invented the electric relay
20	1839	Edmond Becquerel	Discovered the photovoltaic (PV) effect used today in solar panels
21	1857	William Thomson (Lord Kelvin)	Deemed Britain's greatest electrician, invented measuring devices for every electrical quantity in existence and chaired the international committee that named what we know today as the Amp, Volt, Ohm, etc. His home was the first in the world to be lit with electricity and he produced the first mathematics in support of Faraday's concept that electric induction takes place through an intervening medium

	Date	Inventor	Discovery
22	1860	James Clerk Maxwell	His equations unifying magnetism, electricity, and light, make him the second most important scientist in history (next to Sir Isaac Newton)
23	1878	George Armstrong	First hydroelectric plant built in England
24	1879	Thomas Edison	Made the first workable light bulb
25	1882	Thomas Edison	Built the first DC power station in New York
26	1883	Nikola Tesla	Invented the "Tesla Coil," which made it possible to transfer electricity over long distances
27	1884	Nikola Tesla	Invented three-phase alternating current. Electrical power all over the world is based on this principle.
28	1886	Jacob F. Schoelkopf	First commercial hydroelectric power system opened in Niagara Falls, New York
29	1888	Charles F. Brush	Invented an electricity-generating wind turbine
30	1888	Nikola Tesla	Invented the rotating magnetic field
31	1888	Nikola Tesla	Built the first step-up and step-down transformers
32	1888	Nikola Tesla	Designed, patented, and built the first alternating current induction motor
33	1889	Nikola Tesla	Designed and built a system to power electric lights wirelessly at a distance of 30 miles
34	1897	J. J. Thomson	Discovered the electron
35	1904	John Ambrose Fleming	Invented the thermionic vacuum tube (diode)
36	1911	W. Carrier	Invented air conditioning
37	1911	Heike Kamerlingh Onnes	Discovered superconductivity

(*Continued*)

Table 2 *(Continued)*

	Date	Inventor	Discovery
38	1925	Julius Edgar Lilienfeld	Invented, built, and patented a solid-state transistor that used an electric field to control its functions
39	1926	Paul Eisler	Invented the printed circuit board
40	1931	Julius Edgar Lilienfeld	Invented, built, and patented an electrolytic capacitor
41	1931	Soviet Union	First wind energy plant went online
42	1938	Otto Hahn	Discovered nuclear fission, which was the basis of the first atomic bomb (1942) and the first nuclear power plant (1951)

The red arrow in Fig. 46 shows the trend of great electrical discoveries up to 1938: strong and healthy.

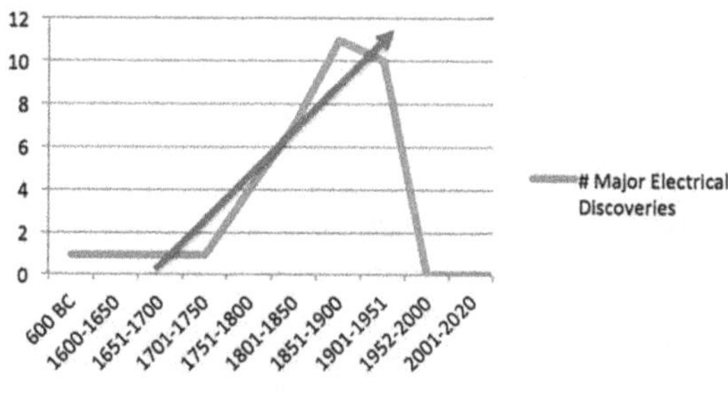

Figure 46 Trend of major electrical discoveries up to 1938.

What happened after 1938?

Why has there not been a basic discovery in the subject of electricity for over 80 years?

If you are confused or incensed by this allegation and pointing to your smartphone, please be patient for a moment. Smartphones are great, but they are only applications of electricity—a way of using electricity. The smartphone is the child of electronics. If you are on a deserted island or on Mars with the latest and greatest iPhone but no source of electricity to charge that phone, you will appreciate quickly enough the difference between electricity and electronics. And there is no need to travel that far. The world's biggest electronics expo, International CES (Consumer Electronics Show), held at the Las Vegas Convention Center in 2018, found itself in darkness for several hours when its electricity supply failed. Hundreds of high-tech power-hungry electronic applications from robot dogs to smart fridges and advanced cooking appliances to sleep aids and speaker systems all suddenly were out of power and out of luck. Even the booths introducing self-driving cars were plunged into darkness. Their technologies were helpless without a source of electrical power. Intel and Sony executives took to Twitter to crack jokes.

There are basic discoveries, and there are refinements of those basics. People like to think (and will be quick to tell you) they have made a "basic discovery" when they have only developed a new application from an earlier discovery. As can be seen from the preceding chart, even many of the basics of the electronics revolution (including diodes, triodes, transformers, transistors, capacitors, and printed circuit boards) were discovered before 1938.

Hydroelectric power is by far the most productive source of green energy. It generates electricity without the pollution of fossil fuel emissions. But its development is owed to the civil and mechanical engineers who fashioned the huge concrete dams that created the vast human-made lakes out of nothing in great projects like the Grand Coulee and Hoover Dams in the U.S. and the Three Gorges in China. The electrical side of hydroelectric has been unremarkable, with the huge, low-speed generators and accompanying switchgear, transformers, and transmission lines following closely on older practices. The world's newest and largest hydroelectric power plant, the Three Gorges in China, still uses the Francis generator, a fully submersible turbine invented in the 1840s by the American James B. Francis.

Wind energy is another example of refinements being touted as basic discovery. In the last few decades, we have built larger wind turbine generators and connected them up to the grid, but the Soviet Union built wind energy plants in the 1930s and the basic idea of using energy from wind was piloted 1000 years ago in China. As for solar, the technology on which our solar panels run dates from 1839.

Yes, they are much more efficient now than they were 100 years ago, but still fall way short of replacing fossil fuels.

Figure 47 One of the 34 Francis generators built into China's Three Gorges Dam.

The last major discovery in electrical energy could be described as how you Tickle the Tiger's Tail by making plutonium cease to exist in atomic fission or making hydrogen disintegrate in atomic fusion. The basic discoveries that led to both the atomic bomb and nuclear power plants occurred in the 1930s.

So how have we been doing over these last 80 years? The red arrow in Fig. 48 shows the electrical discovery trend after 1938. You cannot even call it "unhealthy." It is totally dead. And the question that must be asked is: why?

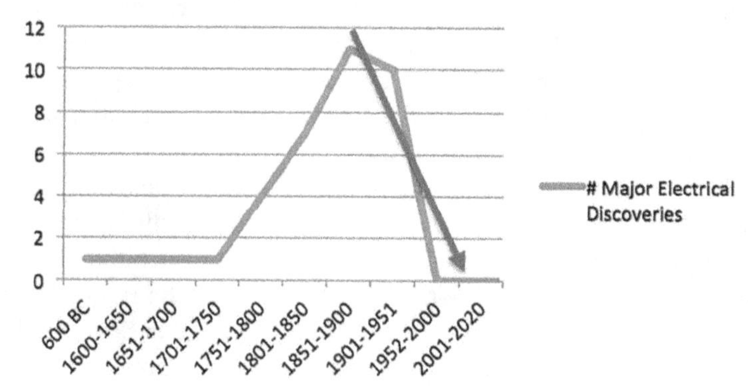

Figure 48 Trend of major electrical energy discoveries: 1938 to the present.

If we want to know why we are tied to a 19th-century power generation technology that is poisoning our planet and most likely also turning the climate against us, we need to determine exactly "what changed." The subject of electricity used to be a vibrant one full of excitement and innovation. So, what killed it?

In the mid-1930s, a new Sheriff came to Science Town. He brought with him a novel approach to the concepts of matter and energy. It was called quantum mechanics and whether you believe it wonderful or whether you believe it not so wonderful, one thing is for certain: **the moment science began to accept its tenets in the 1930s, forward progress in energy technology promptly ceased.**

A good friend of mine, a physicist with a PhD, after reviewing this book, suggested it was wrong to place the 100-year lack of progress in electrical engineering at the door of modern physics. Maybe the older classical physics had run its course by the 1930s, and it was time for a change. One might accept that premise if the results of classical physics had been tapering off for a period of time leading up to the 1930s. But the graphs in Fig. 46 and Fig. 48 tell a different story. Before the classical ideas of energy were abandoned in the 1930s, energy discoveries were wide ranging, frequent, and significant. The exact moment when the concepts of relativity, curved spacetime, and uncertainty took hold, advances in energy discovery ground to a halt.

The effects of global warming are certain to be devastating. To arrest them, we need an Energy Miracle. To get one will require some new viewpoints in science, particularly in the understanding of energy. In the remaining chapters of this book will be found a map to help the science of electrical engineering reorient itself and get back on track.

Thinking outside the electrical box

Tesla's 100-year-old invention of three-phase alternating current was a breakthrough of magnitude in the production and distribution of useful electric power, which is almost universally underappreciated. He developed unique working electrical systems based on the same five basic concepts that had informed all the great electrical pioneers that had come before him, including Franklin, Faraday, and Maxwell. By the mid-1930s, these basic concepts had been largely abandoned.

Figure 49 It is often better to think outside the box.

A time-honored practice for reversing non-optimum situations and restoring positive results is to discover **what changed**. By nullifying a disadvantageous change in course, or restoring a successful one, one can usually get back on track.

In this case, we are looking into the process of how raw electricity gets generated in the first place. And the changes that occurred in the mid-1930s are clear. We need a 21st-century Energy Miracle, and resurrecting the workable theories that were producing Energy Miracles before the 1930s is the logical place to start.

The planet is being poisoned. The Earth is heating up at a rate that is nothing short of alarming. And the world needs to come together in a common cause to find an Energy Miracle.

Robert Lucky's question (posed earlier in this chapter) can now be answered. The electrical engineers will be those who are pursuing and discovering the Energy Miracles.

Chapter 6 reviews the five keys to the Energy Miracles that were abandoned in the 1930s.

Chapter 6

Five Keys to the Energy Miracles

What happened in 1938? Looking closely at the major electrical discoveries in Table 2 in the last chapter discloses five electrical fundamentals that informed almost every major electrical advance on that chart. Amazingly, every one of those five was discarded by the mid-1930s. And it can be accurately said that when these key functions were abandoned, advances in new electrical power generation ceased.

Vince Lombardi, the legendary football coach whose successes made him a national symbol of the determination to win, always started his team's training camp the same way. He would hold up a football so it could be seen by all his players (who were, by the way, the most professional and experienced footballers in the world), and then he would say with great enthusiasm, "This is a football." The point he was making was that we begin with basics because only the basics make us champions. This is all the more true in energy science.

In the interest of starting a dialog that will turn the switch of Energy Miracles back on, the five major electrical basics that informed every one of the world's major electrical breakthroughs will be discussed in this chapter. They include:

1. The very definition of energy itself,
2. What gives energy its impetus,

Energy Miracles: The Global Warming Backup Plan
H. B. Glushakow

ISBN 978-981-4968-18-8 (Hardcover), 978-1-003-28442-0 (eBook)
www.jennystanford.com

3. The actual structure required before energy can occur,
4. The requirement of a medium for energy to propagate through, and
5. The actual mechanism by which energy propagates through that medium.

Key #1. Energy: What Is It?

"If you want to find the secrets of the universe, think in terms of energy...."

—Nikola Tesla

These days when you ask the question, "What exactly is energy?" you receive answers such as:

"It's the total energy of a system."

"It's the composite of an object's kinetic energy and potential energy."

"It's made up of kinetic energy, gravitational energy, and thermal energy."

"It's energy because it's conserved. It transfers between objects and it never changes. As objects move around, the amount of energy associated with them never changes."

In physics, the term conservation refers to something that does not change, and energy is considered to be a conserved quantity. But that does not define energy, and this is too important a word to lack a definition. The most honest answer is the admission given by Nobel Laureate Richard Feynman half a century ago: "It is important to realize that in physics today, we have no knowledge of what energy is."[29]

The elusive definition

The aspiring seeker of Energy Miracles, starting his journey bright and bushy tailed, must be prepared for some nasty shocks. The first thing he will find is how poorly the basic concepts of energy science

[29]R. Feynman, Conservation of Energy, in *Feynman Lectures on Physics*, California Institute of Technology, 1963.

are defined. Thomas Young first introduced the term "energy" to the field of physics in 1800 in explaining his experiments involving particles of light, moving through a medium and thus causing waves. He was talking about things moving or changing as the result of the impact of actual physical particles. This was corroborated and built upon by the great electrical discoverers of the 19th century: James Prescott Joule, Hermann von Helmholtz, William Thomson (Lord Kelvin), Michael Faraday, James Clerk Maxwell, and others. Somewhere along the way, this concept got lost.

The *Longman Science and Technology Dictionary* now defines energy as "**The capacity of a body for doing work**" as does the *Encyclopedia Britannica* and almost every physics textbook. Though very neat and easy to remember, that definition affords very little assistance to someone in search of new energy sources. It may tell you what energy is *potentially capable of doing*, but it is certainly not a description of what energy is.

There are various kinds of definitions. The most common type is that which classifies a thing by describing its characteristics. One can also define something by comparing its differences or similarities to other things. The best type is the action definition, which notes how something comes into being, changes, and ceases to exist. This last type is what is needed for progress to be made in energy discovery, and the preceding definition in boldface type does not even come close to it. A clear definition of energy is necessary to overcome many of the problems blocking the discovery of Energy Miracles.

"The capacity of a body for doing work" is not a very scientific definition. It is at best a vague characterization. It does not tell us what energy is or how it is generated or how it can be measured.

Figure 50 According to modern physics, energy is a mysterious characteristic of matter that leaves and arrives higgledy-piggledy.

Yet, when it comes down to it, most of us do have a pretty good idea of what energy is. We have all looked up at the sky to see nature's best show and demonstration of energy: lightning. Anyone who has ever been at or near the receiving end of a lightning strike knows how much energy is wrapped up in one of them. The huge explosion it causes can be heard and felt for many miles. The mechanism behind this phenomenon is explored in the last chapter of this book. But we can simplify it here as *the flow of electrical particles from the clouds to the Earth.* We can all agree there is lots of energy there. And of interest is the fact that we do not usually think in terms of lightning "having" energy. Rather, we say lightning **is** energy.

Why this is important is that following right along from the definitions of energy cited earlier, those same physics textbooks describe electricity as *"an electric charge that lets work be accomplished."* With respect, this is just not very useful to someone in search of an Energy Miracle who is seeking to wrap his or her wits around the subject of electricity. When I first studied electricity in the 1960s, one of my first textbooks stated that no one really understood electricity and the most the student could achieve was to learn enough to handle it safely. That was nonsense then (as I found out when I got a bad electric shock perched atop a 25-foot ladder), and it is nonsense now.

The actual definition

Energy is not a spooky thing that allows something else to create an effect. Energy is the thing doing the affecting. **Energy** is nothing more, nor less, than **postulated particles in space.**[30] The particles can be of any size: large as ping-pong balls, railroad cars, or planets; or small as atoms, electrons, or photons. In physics, *postulated* is a specially applied word with a causative or dynamic connotation. It implies a motion of starting, stopping, or changing, which can set a pattern for the future or nullify a pattern of the past. Particles are normally called energy when they flow from one point in space to another. They can create a direct impact as in the top frame of Fig. 51, or they can transmit an impulse as shown in the device in the bottom frame, known as Newton's Cradle:

[30]This concept was respected by Ampère and Faraday, foundational for Maxwell and Lord Kelvin, codified by Hubbard, and employed to great effect by Nikola Tesla.

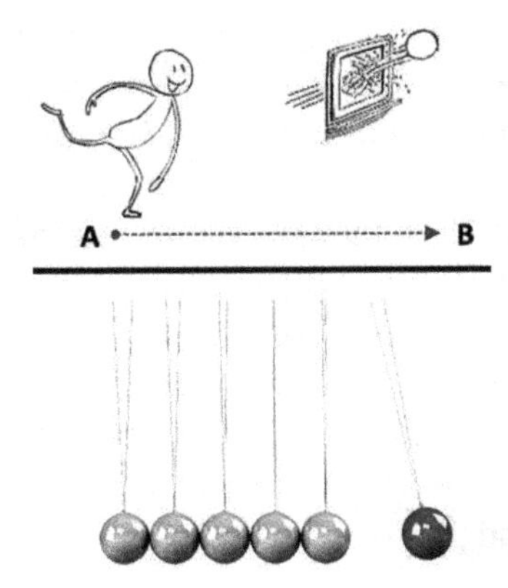

Figure 51 Two effects of energy: Top frame: particle traveling from Point A to Point B. Bottom frame: Newton's cradle showing an impulse being transmitted.

The concept of energy became muddied in the early 1900s as a result of several confusions, the main ones being: Was "light" the same kind of energy as "electricity" or "atomic"? What was energy made of? Was it a particle or a wave? Could it be created or destroyed? Did it need a medium through which to propagate? If so, what was the mechanism of that propagation?

We shall deal with each of those facets in detail in the second part of this book, but suffice it to say here that every major electrical discovery was premised on the fact that energy consisted of actual particles moving through some type of medium, whether a solid copper wire or in free space manifesting wave-like qualities.

Energy and its conservation

Before the dust has settled on the aspiring seeker of Energy Miracles who has discovered his textbooks do not define energy very well, he is immediately assaulted with the exhortation: "ENERGY CAN NEVER BE CREATED!" This comes from the law of the conservation of energy, which says energy can never be created or destroyed. We do not criticize this law because there are many applications for it. The problem with it is not so much the law itself, but the fact that quantum mechanics has made it pretty much the sole definition and criterion of energy, which it plainly is not.

Imagine a law of conservation of cats that says cats can never be created nor destroyed. The total quantity of cats is conserved and must remain unchanged. So if you had two rooms and two cats in one of those rooms, you would know from this law that if some or all of the cats disappeared from room 1, you could be sure to find it in room 2. (That is, they could never be created or destroyed, just moved around a bit.) The thing is, nothing in that description tells you the first thing about what is a cat.

We are not challenging whether or not energy is conserved. We are just making the point that, at best, this is just a characteristic of energy, not a complete definition. Because quantum mechanics has injected so many twists into the subject of energy, we take it up further in the second part of this book.

Key #2. Dichotomy: The Condition That Gives Energy Its Impetus

The word **dichotomy** comes from two Greek words meaning to cut into two parts; and that was exactly its original meaning. Today it is generally thought of as two things opposed to each other that are entirely different. This is not the case on two counts:

1. The two parts of a dichotomy cannot be *entirely* different. They have to have something in common to be a dichotomy. Good and bad may be a dichotomy when they are both describing the behavior of a child. Good behavior and bad weather are not a dichotomy as they have nothing in common.
2. Although a dichotomy may be stated as a pair of two, there is always a range between those two extremes. You could have an extremely bad boy, a moderately bad boy, a bad boy only some of the time, a good boy, a very good boy, and a boy of exceptional behavior. There is an infinite range of potential differences between the two parts of a dichotomy.

In physics, a **dichotomy is a pair of opposites which when interplayed cause action**.

All energy production comes about from this mechanism. Maxwell's equations prove this for both electricity and magnetism, but as I promised not to include mathematical equations in the book, you will have to take my word for this.[31]

---

[31]Well, for those who are really interested, the equations can be found in the Mathematics Postscript.

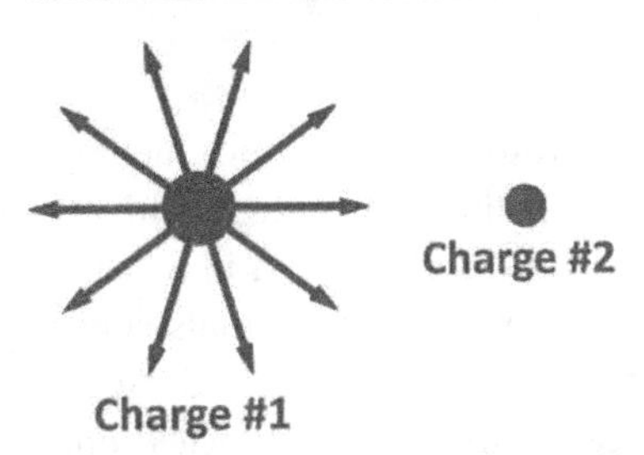

Figure 52 Two points are always needed to create energy.

Physics texts (such as Tipler) teach "single-point charges" (see Fig. 52 Charge #1). However, energy requires a difference in potential between two terminals, so a second charged point (Charge #2 in the figure) is always needed for electrical energy creation.

Figure 53 shows common dichotomies known to produce energy. There are other dichotomies that could also be made to do this and sometimes (as in magnetism) when you bring two pluses together, you can also get a flow of energy.

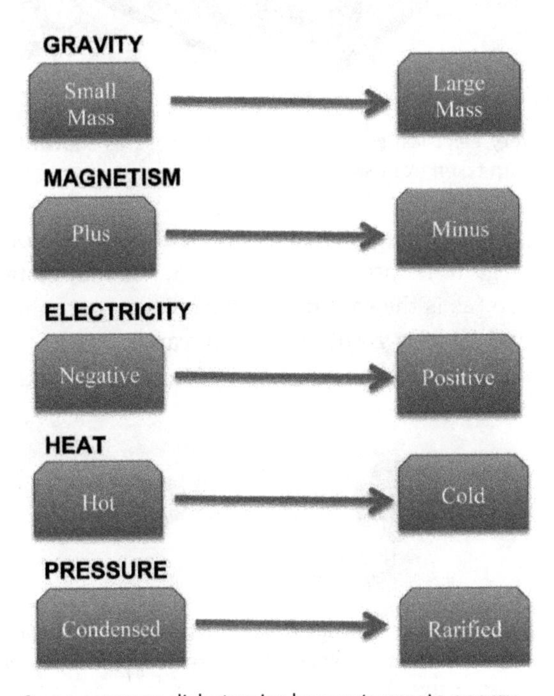

Figure 53 Some common dichotomies known to create energy.

Key #3. Structure Required before Energy Can Occur

In any electrical generator or energy-producing device, a base must be established to keep the two poles/terminals of the dichotomy apart. See Fig. 54. This is easily seen in a battery where the base keeps the positive and negative terminals separated, but it is also true for any motor, generator, or other device where you want to encourage or create the motion of particles (energy) and is proven by Coulomb's law.[32]

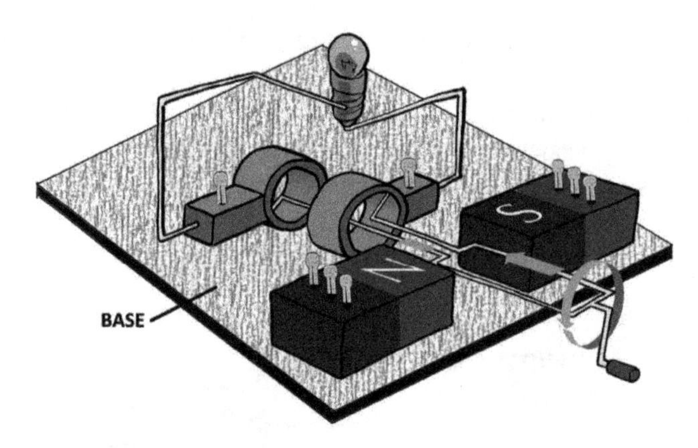

Figure 54 Every electrical generator requires a base as shown here in this modified diagram taken from Siyavula.

Distance must be maintained between at least two objects for electrical energy to be produced. In Fig. 55, *r* is that distance, and if *r* equals zero (as is the case of a single object where the distance between it and itself is zero), no energy can be created. This is a vital element to consider for new Energy Miracles, for without it, no energy could flow.

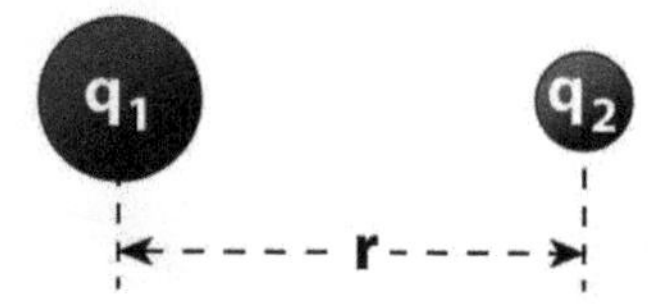

Figure 55 Space between two objects with charges of different potentials must be maintained for electrical energy to be created. (The "*r*" is the distance between those two charges.)

[32]For formula, see Mathematics Postscript.

Key #4. Medium Needed for Energy Propagation

Some sort of medium is required through which energy can propagate. This holds true for energy flows of every speed and wavelength on the complete scale of electromagnetic radiation. A medium can be a solid material such as a copper wire or a liquid electrolyte. It can be a stream of ions created and then used by a beam of light to propagate itself, or it can be some form of an aether medium. (Yes, I used the term aether knowing full well it is the third rail of science and only thinking the word has been known to burn an academic to a crisp.) I am not now, nor have I ever been, a member of the Aetherist Party, yet it must be said that those who used this concept in their research produced some amazing Energy Miracles, and those who abandoned it, produced nothing.

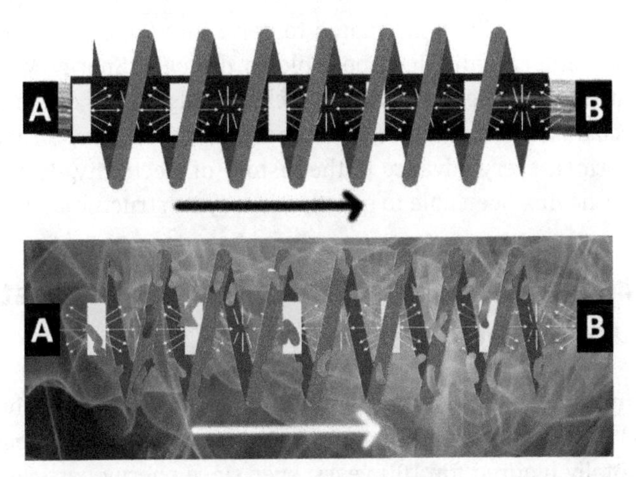

Figure 56 Energy flowing through different media. Top frame: through a copper wire. Bottom frame: through free space. Faraday, Maxwell, Tesla, and others concluded the mechanism was the same in both cases.

Sir Isaac Newton, the greatest scientist in history, explained it thusly, "It is inconceivable that inanimate matter should, without the mediation of something else which is not material, operate upon and affect other matter, without mutual contact. That...one body may act upon another at a distance, through a vacuum, without the mediation of anything else, by and through which their action and force may be conveyed from one to another, is to me so great an absurdity, that I believe no man who has in philosophical matters a competent faculty of thinking can ever fall into it."[33]

James Clerk Maxwell, the second greatest scientist in history, based his famous equations on it, saying, "Whenever energy is transmitted from one body to another, there must be a medium or substance in which the energy exists after it leaves one body and before it reaches the other...it is inconceivable that a wave motion should propagate in empty space."[34]

Nikola Tesla (equally brilliant in his own way) said, "All attempts to explain the workings of the universe without recognizing the existence of the aether and the indispensable function it plays, are futile and destined to oblivion."[35] Tesla used the Earth itself as a medium for conducting electrical current over long distances. Even Einstein accepted the aether until he did not.

We shall introduce the aether medium in detail in Chapter 8; and while some will be fascinated to delve into the exact structure of these aether media, it is possible to discover Energy Miracles without that knowledge simply by acknowledging the existence of the mechanism. Remember, some sort of medium (such as an aether) has informed every advance in the history of electricity, despite no one ever having been able to exactly identify its structure.

Key #5. Mechanism by Which Energy Propagates through the Medium

Knowing how electrons or photons move through a medium is critical for anyone attempting energy research, yet the subject has been totally ignored for 100 years, ever since energy particles and mediums were both deemed "imaginary" by quantum mechanics. Textbooks from Franklin's time all the way up through the first quarter of the 20th century explained in clear and straightforward terms that energy propagated in waves through a medium with little differentiation made between light waves and sound waves. For example, the interferences (such as diffraction and refraction)

---

[33]Sir Isaac Newton, 1692, third letter to Bentley.=

[34]James Clerk Maxwell, *A Treatise on Electricity and Magnetism*, Vol. 2, last page, Oxford University Press, Oxford, England, 1892.

[35]N. Tesla in a prepared statement on his 81st birthday on July 10, 1937.

affecting both wave types were explained identically.[36] Four more such textbooks are cited in this footnote.[37] The Nobel Laureate Richard Feynman also showed that electromagnetic waves operate by the same wave theories as does sound.[38]

All the greatest electrical discoveries were premised on the fact that energy of all types (including electricity and light) propagated through some kind of medium. And then one day, in the early 1900s, light did not. What made light waves different? Einstein did. This is dealt with in more depth in Chapter 8, but what we find here is the abandonment of the greatest electrical discovery in the history of the world: Maxwell's theory of electrodynamics. It also pretty much closed the door on Energy Miracles.

Figure 57 is a classic condensation–rarefaction wave, still used in other realms of science, but abandoned in the fields of electromagnetics and energy research ever since quantum mechanics threw it out along with the precipitous decision to banish the wave theory of electromagnetic waves.

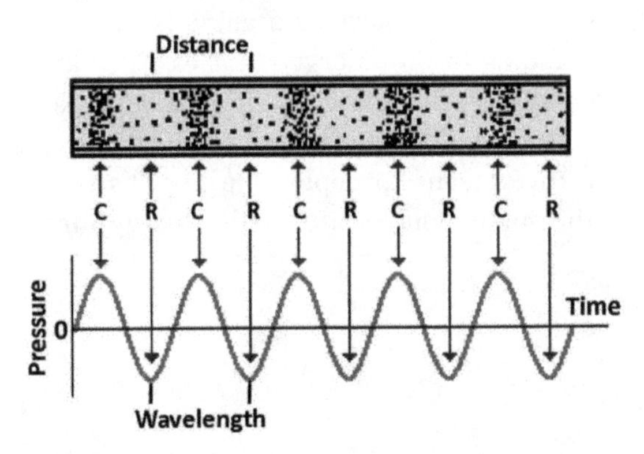

Figure 57 Condensation/rarefaction wave. The "C"s are condensations (periods of greater pressure) and the "R"s are rarefactions (periods of lesser pressure).

[36] J. A. Culler, *A Text Book of General Physics: Electricity, Electromagnetic Waves, and Sound*, J. B. Lippincott Company, Philadelphia, 1914, p.272.

[37] A. W. Smith, *The Elements of Applied Physics*, McGraw-Hill Book Company, New York, 1923, p.250; R. J. Stephenson, *Exploring in Physics*, University of Chicago Press, Chicago, Illinois, 1935, p.96; A. P. Gage, *The Principles of Physics*, Ginn and Company, Boston, 1895, p.267; W. S. Franklin and B. Macnutt, *General Physics*, McGraw Hill Book Company, New York, 1916.

[38] R. Feynman, *Feynman Lectures on Physics*, #47 Waves.

In this figure, all the parts of an electromagnetic wave can be seen. The dark bands are condensed particles (electrons or photons). In between those condensations, you can see the areas of rarefied (less dense) particles. The wavelength is the distance between two condensations (or two rarefactions). The frequency is how many of those wavelength cycles occur in a given period of time, such as every second.

When an electrical current flows through a medium, the individual electrons do not move very far. What happens is that particles push out from one terminal of the dichotomy in the direction of the other terminal, causing the particles ahead of it to condense, creating a relative solidity. At a certain point, the condensation changes polarity and explodes outward, pushing other particles away from it (rarefaction) in the direction of current flow toward the opposite terminal of the dichotomy. In such a series of condensations and rarefactions, the electrical current moves rapidly through the medium.

Using the previous five keys as a guide is the path of golden breadcrumbs leading to the Energy Miracles. They are each dealt with in more detail in the second part of this book, but in the next chapter we introduce the Energy Miracle Challenge, a political and administrative means to deploy the planet's resources most efficiently in the pursuit and location of the Energy Miracles.

Chapter 7

The Energy Miracle Challenge: Solution to Global Warming

Folks both cruel and charming,
Let ancient discord cease.
From countries old and warming
Awake the song of peace.
The seas they may be rising
While energy's embalmed
But science enterprising
Can render earth encalmed.

The Glory Days of Electrification

The U.S. and Europe experienced a rapid period of industrialization between 1760 and 1820. Modernized by steam power, industrial processes began switching from hand production to machines with great benefit to the societies concerned. Between 1870 and 1914, a second Industrial Revolution occurred, marked by even greater advancements in manufacturing and production. Its greatest achievement was electrification. For the first time electrical power was available to homes and factories while public utilities like streetlights were being introduced in cities across the world. Electric lighting improved working conditions and AC electric motors

Energy Miracles: The Global Warming Backup Plan
H. B. Glushakow

ISBN 978-981-4968-18-8 (Hardcover), 978-1-003-28442-0 (eBook)
www.jennystanford.com

reduced production costs in many industries and made possible the production assembly lines that became the mark of the times. Electric trains improved travel and accessibility. Millions of jobs were created. It was the sudden abundance of cheap energy that enabled the growth of industrialization, urbanization, and globalization that has dominated the world for the last 150 years.

The initial rollout was speedy and impressive. Three years after the invention of an electric motor that could run them, 110 railway lines were all using electricity. In the U.S. the first large-scale power station went on line in New York in 1882 and within 3 years a handful of others were completed, serving several hundred customers. Eight years later there were 1,000 central stations in operation serving many thousands of customers in the U.S. By 1902 the number had jumped to 3620, and by 1925 half of all power was being provided by centralized electric power stations. At the turn of the century, steam power was driving 80% of America's industrial machines. Twenty years later 80% of those machines had been electrified.

None of this happened by coincidence. This expansion was accomplished by a combination of and coordination between government, corporate America, and institutions of higher learning. Congress passed laws to enable this expansion. The tens of thousands of trained personnel that were needed immediately to run these machines got trained, many by universities and many others by the corporations themselves. Large, well-organized research laboratories, many of which engaged in in-house training, appeared in companies like General Electric (1900), Westinghouse (1903), American Telephone and Telegraph (1913), Bell Telephone (1913), DuPont (1911), Eastman-Kodak (1912), Goodyear (1908), General Motors (1911), U.S. Steel (1920), and Union Carbide (1921). One way or another all of them focused on electricity. By the 1930s there were over 1600 firms in the U.S. with research units. And many of them recruited university laboratories as quasi-annexes. The universities were all too happy to cooperate. It was the heyday of scientific advance and innovation.

The enormous costs of some of those 20th century projects encouraged partnerships: both domestic and foreign. The Hubble Space Telescope, a NASA project in partnership with the European Space Agency, is an example.

Figure 58 NASA Hubble Space Telescope.

It was conceived in the mid 1900s and took many thousands of people to build it at a cost of over $2 billion making it the most expensive scientific instrument ever constructed. It was finally launched into space in 1990. Any such feat, and this one in particular, was only possible by combining the forces of many governments, companies, scientists, and individual workers. It is exactly this kind of determination and cooperation that is required right now to solve our looming energy crisis.

We need to replicate that motion today, and this chapter proposes a way to do just that.

Strengthening the Paris Climate Accords

Now that the U.S. has rejoined the Paris Climate Accords, the world is pretty much in 100% agreement that the effects of global warming need to be confronted and mitigated on an immediate basis. Only by working together can this be done. Every step already taken or planned by countries to reduce carbon emissions with wind, solar, or hydro projects should be reinforced and augmented.

Dealing with the most serious consequence of climate change

There are many things that solving global warming will not fix. It will not fix husbands cheating on their wives or wives cheating on their husbands. Neither will it cure cancer, or dogs ruining the good furniture. It may be that more global warming is unavoidable, with some terrible consequences down the road. But those consequences need not be fatal. With the willingness of the world to come together and confront it, we can yet get a handle on the effects of Global Warming.

A big question is what would be the worst consequence of the planet's temperature increasing by 2 or 3 degrees in the next few years? With all due respect to those beautiful animals, it won't be the death of a few hundred polar bears or even their complete extinction. Neither will it be an extra super storm or two per year or more forest fires. Water and food will become even scarcer, which will foment unrest and social chaos across the world, but that will not be our biggest problem.

The planet's biggest problem will be how to provide for the relocation of the one billion people currently living in low-lying coastal areas; because the seas are going to rise. We are not just talking about putting your family into a car and driving 50 miles inland. We are talking about transplanting huge cities with millions of inhabitants; and rebuilding the accompanying housing as well as the commercial and industrial infrastructure. Directly or indirectly every continent and every country will be heavily impacted by this event. Every U.S. state will bear the brunt of this including landlocked and coal-producing areas like Wyoming, Illinois, West Virginia, Pennsylvania, and Kentucky. It will require tremendous amounts of energy for the moving and transportation; and even more energy for the construction of new infrastructure—roads and bridges, Internet and telecommunication networks, plus housing and industrial plants. Many of the very electrical power facilities that would be called upon to provide the huge amounts of electricity this will entail will also be swamped by the rising tides. This is a giant undertaking and one that requires we come up with one or more new energy sources.

A not always obvious fact is that if you exactly name what it is you want, and if you really want it, and if you then make a determined effort to figure out how to get it, you are likely to succeed in getting it. It is so with alternative renewable energies.

The successful pursuit of new power sources that gets the planet to a net Zero of carbon emissions is the ultimate goal of

this game. But that goal may be too broad, so we suggest, as the first point of attack, the replacing of all electricity now being generated by fossil fuels.

Electricity from Fossil Fuels: Primary Source of Greenhouse Gas Emissions

We start with electricity because with that single action, we'll eliminate 50% or more of yearly greenhouse gas emissions. That's a reduction of over 26 billion tons out of the total 51 billion per year we are now emitting globally. The reasons this is a good first target include:

- It directly tackles the impact of coal. Believe it or not, despite all the promotion of wind and solar technologies, in recent years coal has been the world's fastest-growing energy source by quantity. Coal accounts for almost 40% of global electricity production making up 50% of U.S. electricity production, 60% of electricity production in China, 70% in India, and over 85% in South Africa. At the same time coal has the worst environmental impact of any of the conventional fossil fuels due to the damaging impacts of both digging it out of the ground and then burning it to release its energy. Although it contributes less energy to the world economy than petroleum does, coal is the primary source of greenhouse gas emissions, producing over 40% of the world's CO_2 emissions. (94 kg of CO_2 are emitted for every gigajoule of energy produced from coal.)
- Replacing fossil fuel–generated electricity with a clean, cheap, and reliable energy source not derived from burning oil, coal, or gas will reap benefits outside of the traditional uses of electricity. At this writing, electric cars are within our reach. We can soon be rolling them out in quantity to replace gas-burning vehicles. But without a replacement for fossil fuel electricity to keep their batteries charged, almost all the electricity that will be used by these electric vehicles will come from coal-burning generators.
- Consumers who are currently using fossil fuels in other industrial and heating applications will be able to replace them with the availability of a new source of electricity.

Figure 59 is a conservative estimate of the change in the ratio of fossil fuels to clean alternative fuels from the single action of eliminating electricity from fossil fuels.

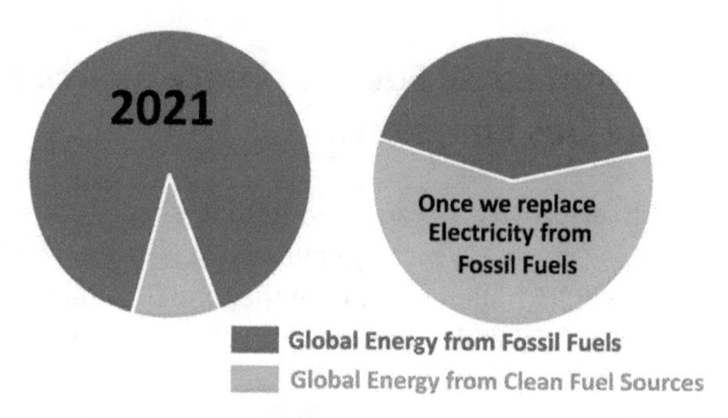

Figure 59 Left circle: Today's % of global energy from fossil fuels (red); Right circle: The change created by the single action of eliminating electricity from fossil fuels.

What we are discussing here is a revolution in electricity production initiated by the discovery of an Energy Miracle—an available, clean, reliable, and cheap source of electricity. Once such a system of energy is discovered or invented, the great many uses to which it will immediately be capable of being put will greatly simplify the job of rolling it out at the scale needed for meaningful global impact. Said another way, it is envisioned that this will be a "one size fits most" operation.

There are other uses for fossil fuels than generating electricity which will not be ameliorated by a new source of electricity. Coal creates CO_2 emissions in the production of steel and the manufacture of cement. In many countries it is still a major source of heating. The same holds true for oil and natural gas. A new source of electricity will not directly impact the emissions created by agricultural (including livestock), long-distance transportation (including long-haul trucks, ships and planes), or in many other areas of industry. However, the simple replacement of fossil fuel electricity with an Energy Miracle will get us more than half-way to our goal in one fell swoop while solving many of the other problems associated with fossil fuels such as water and energy scarcity, food shortages, and pollution.

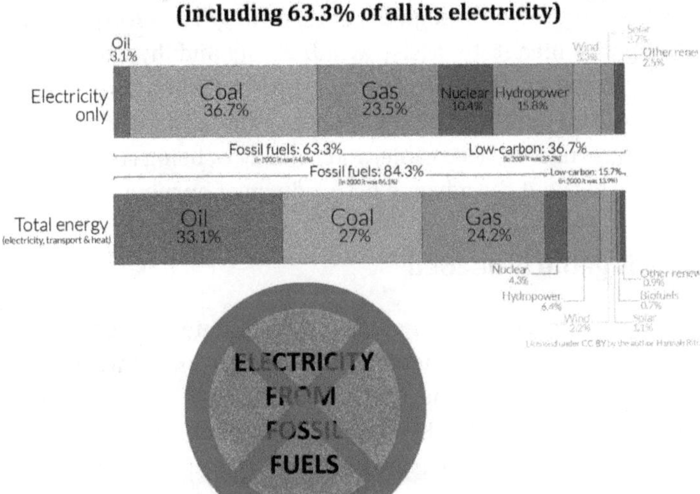

Figure 60 Stamp out electricity from fossil fuels.

The Energy Miracle Challenge

"Energy innovation is not a nationalistic game."

—Bill Gates

"Only a free individual can make a discovery...the elimination of independent groups leads to one-sidedness and barrenness in science."[39]

—Albert Einstein

The **Energy Miracle Challenge** is an undertaking in the form of a game intended to solve the world's energy/climate crisis. It is premised on several facts, documented earlier in this book, that fossil fuels are both poisoning and warming the planet, but even if they weren't, they are inadequate to satisfy even the most basic energy needs of the planet. At this writing in the fall of 2021 there's not enough coal in China and India, not enough gas in Europe, AND

[39]Einstein, Albert. *Ideas and Opinions*, Crown Publishers, New York, 1954.

not enough oil everywhere, including in the U.S. where gasoline prices have exploded through the glass ceiling. The **Energy Miracle Challenge** is meant to assist wind, solar, and hydro power in replacing fossil fuels by discovering another source of electrical energy—a 21st Century Energy Miracle—that is at once renewable, cheaper, dependable, accessible, and capable of being rolled out on a large scale without the adverse by-products of burning fossil fuels.

Why this game is needed

Global action to confront the effects of climate change began in 1992 when the U.N. Framework Convention on Climate Change (UNFCCC) promulgated the first climate change treaty, signed by 154 countries. In 1997 the Kyoto Protocol provided the first attempt at implementing actual measures to achieve the goals of that treaty. This was superseded by the Paris Agreement of 2015 which by 2021 had been ratified by almost 200 countries. These actions were done to bring the world together and arrive at a coordinated consensus on how best to deal with the challenges of global warming.

The Conference of Parties (COP) is the U.N. decision-making body directly overseeing global climate change activities. This body (COP) is arguably the most important political force in the world today. For one thing, it is leading both the attack and defense against the world's biggest threat. And for another, it speaks on behalf of **all** the peoples of Earth, the first time anyone or anything has done so in the 10,000-year history of the planet.

Its 26th annual meeting (COP26) was held in Glasgow, Scotland in November 2021 and it is unfortunately the case that the climate change treaty movement may be running out of gas. While global awareness of the climate change threat is widespread and increasing, major climate targets are passing by unmet, and some of the earlier enthusiasm has been dulled. This can be seen in the following resolutions that came out of COP26, taken from the official papers posted on its website:

(1) **Assistance to Developing Countries.** The Paris Agreement confirmed an earlier resolution that the developed nations would help the 120 less developed ones to tackle global warming in the amount of $100 billion per year in grants

or investment. It was clear from the inception of the first climate treaty that progress in those countries could not be made without such assistance. Over half of these countries are currently suffering so badly from social and economic crises that their ability to make the minimum payments on existing debt is in question, much less undertake new climate initiatives. Yet up to today, 25 years after the original promise, that figure has never been reached, and the unpaid / overdue part of the promised yearly 100 billion must now be approaching one trillion dollars. Achim Steiner, Administrator of the U.N. Development Program, estimated in Nov. 2021 that trillions of dollars would be needed for these countries to begin to meet their targets under the Paris Treaty, and he admitted he had no idea where such funds would come from. Clearly, these unfortunate countries are not going to be transitioning from coal to wind turbines or solar power any time soon—certainly not in the next decade.

(2) **Just Transition.** This was a point of contention in COP26 but it need not be so. In our rush to reduce carbon emissions and slow down global warming, we have to keep in mind that our goal in all this is **to raise** the survival level of the peoples of Earth, not reduce it. There are hundreds of millions of people across the planet who are right now directly dependent on electricity from fossil fuels for their jobs, their ability to feed their families, heat their homes, travel, and for many other things. The poorer the community, the more negatively a transition away from fossil fuels can impact that community, and it is little wonder that many are in terror of a poorly thought-through transition. A vital element for any program of climate change mitigation is provision for those who will be directly oppressed by the loss of fossil fuels and related jobs in their communities. The new term for this is *Just Transition*, called that by those who wish to remind us that the richest countries are also those who over the past century have emitted the most carbon pollution and for the most part continue to do so. Though a true fact, there's no use regretting the Industrial Revolution since every country, rich and poor, has benefited from it. The Energy Miracle Challenge takes this into account by providing the rights to the Energy Miracles

free of charge to all countries rich and poor. I'd prefer to define a *Just Transition* as a transition away from fossil fuels that respects and renders aid to coal-mine workers and others whose jobs will be lost because of it. The best solutions so far put forward by experts and think tanks are based on two facts: there will be lots of people looking for jobs in some communities; and there will be lots of potentially valuable land, much of it currently ruined by coal mining, that could be cleaned up and repurposed for things such as agriculture, fisheries, or tourism. Putting those resources together is a good first step, but will probably require additional actions.

Where possible, coal-generated electricity should be immediately replaced with alternative energy, but precipitously shutting down coal mines that would create dire consequences for local populations need not and should not be done. The search, discovery, and roll-out of Energy Miracles can be done in parallel. The creation of an Energy Miracle will change everything. With a cheap, clean, and efficient source of energy, every country will be able to utilize its unique features to create wealth undreamed of before. Where there has been scarcity, there can be abundance. Where there was no investment available before, there will be many new sources of funds. If your land has nothing but coal mines and deserts, with an Energy Miracle it can be restored to a lush and productive condition capable of supporting agriculture and many other things. Such wealth will guarantee jobs and livelihoods for everyone currently dependent on coal.

(3) **Coal.** The COP26 statement "Global Coal to Clean Power Transition" recognizes that coal power generation is the single biggest cause of global temperature increases and makes some ambitious plans to eliminate coal-powered power generation. The good news is that over 40 countries signed it. The bad news is that the world's biggest coal consumers didn't, including the United States, China, Australia, India, South Africa, Russia, Brazil, and Argentina.

(4) **Global Grid.** The COP26 statement, "One Sun One World One Grid" introduces the Green Grids Initiative. This is an expensive plan to construct a global power grid of electrical transmission lines crossing frontiers and connecting different

time zones, "creating a global ecosystem of interconnected renewable power for mutual benefit and global sustainability." It's a great idea and the planet needed it yesterday, but as far as mitigating global warming, this is putting the cart hundreds of miles before the horse. It's as if you were constructing 500,000 kilometers of modern 8-lane highways with only 5 or 6 cars available to travel on them. Once we've got our energy miracle, such a grid will become indispensable, but our focus must first be on achieving the energy miracle itself.

(5) **Joint Action Necessary**. The COP26 statement on "International Public Support for the Clean Energy Transition" states that "joint action is necessary to ensure the world is on an ambitious, clearly defined pathway towards zero emissions..." The statement "undertakes to end new direct public support for the international unabated fossil fuel energy sector by the end of 2022." There are three problems with it: (a) only 26 countries signed it; (b) It includes a vague caveat that would allow the signatories (few enough though they are) to simply ignore it; and (c) Not a single oil, petroleum, or gas company has signed onto it. In fact, the top 5 oil companies are reported to spend a total of $200 million every year in attempts to block the efforts of COP26. The relative shares of their contributions are shown in Fig. 61.

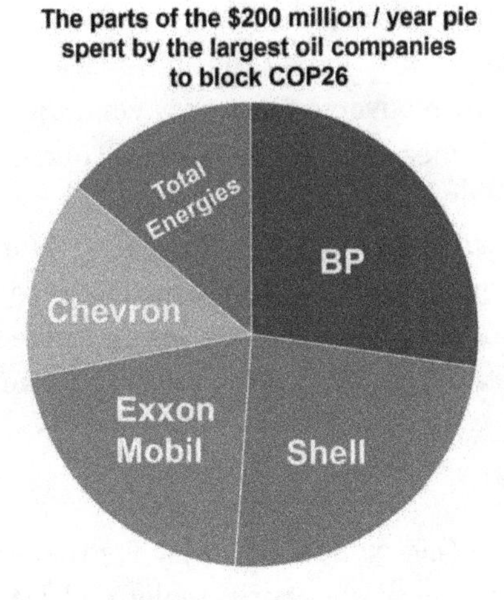

Figure 61 Oil companies spend $200 million each year to block the efforts of COP26. *Source*: InfluenceMap.

(6) **Financing.** Global warming poses a threat several orders of magnitude greater than Covid-19, yet the richer countries hardly blinked at spending 14 trillion dollars in a single year to fight Covid. Not only that, for the last 15 years governments have subsidized fossil fuels to the tune of $423 billion per year and they continue to do so. Yet no money to combat climate change? Why is that? Any apparent apathy at implementing the climate treaty may not stem from a lack of appreciation of the global warming threat. Rather, it may reflect a growing realization that throwing trillions of dollars into existing programs is not going to solve the problem. Many of the developed countries have been active at working on their own climate change transitions, but as we've seen in Chapter 3, overall progress even there has been lackluster. What we need is an Energy Miracle.

(7) **Beating a dead horse.** We know quite well, from experience on the plains of ancient Mongolia to the grasslands of the more recent American wild west, that beating dead horses will not bring them back to life. The "dead horse" alluded to here is NOT the Climate Treaty—it is the unjustified hope that trying to turn wind turbines, solar panels, and carbon capture into the total solution to global warming, is ill-conceived. COP26 ended up kicking the Climate Change Movement down the road for another year, mandating another meeting in 2022 where it is hoped that obstacles such as the above can somehow be overcome. This is what the **Energy Miracle Challenge** is meant to remediate. It will not need anything like the ten trillion dollars discussed at COP26, but it will mobilize all the available resources across the world in countries rich and poor, pointing them in a single direction; putting the brains, available finance, and attention where they will do the most good and actually solve the global warming challenge.

Who can play?

Anyone can play and everyone should play. For the reasons explained earlier, almost no one in the world really understands energy or electricity but with the help of this book, almost anyone could learn.

Today electricity-innovation-and-discovery is a level playing field where anyone with intelligence, skill, and a few instruments can make major contributions. There's not a government in the world that could not afford to enter this game. It doesn't require a billion-dollar Hadron collider. It doesn't require multi-million-dollar fancy research buildings and laboratories. It requires some interested men and women to put their minds to solving the most important challenge of our time: the understanding and harnessing of energy.

It doesn't even require a government to play. Any company or organization can enter. A school can enter. Single individuals can enter, or if preferred, several can jointly enter, and work together as a team. Governments can invest in corporations to enter. Corporations could partner with institutions. The winner simply must actually solve the problem, which is initially to find an energy source that can replace electricity from fossil fuels.

There is absolutely no restriction as to gender, age, race, religion, national origin, continental or cultural heritage, or sexual preference. An unmarried 16-year old green-skinned mother of three is as welcome as an octogenarian dwarf with seven wives.

The rules

The research can be done in every country and in every language, but it is suggested that a single language like English be adopted so that all relevant papers and translated materials can be shared and made available to all.

1. The primary goal of the game is the discovery of new sources of energy that are clean, efficient, cheap and can be implemented at large scales. The first priority is "replacing electricity from fossil fuels."
2. The secondary goal of the game is the creation of novel non-fuel uses for coal, oil, and gas for countries and communities whose economies have depended on the production, refining, or sales of these three substances. Fossil fuels can be consumed, but not combusted, when they are used directly such as for construction materials, chemical raw materials, lubricants, solvents, waxes, and other products. In 2017, seven percent of all fossil fuels were consumed for non-combustion

uses in the U.S. That includes thirteen percent of all petroleum products. More uses can be found. Although an important part of the Energy Miracle Challenge, this provision alone will not necessarily guarantee a Just Transition. (More on Just Transitions can be found earlier in this section.)

3. The Secretary General of the United Nations will appoint a 3-person panel from a list of 10 candidates proposed to him by a vote of all signatories to the Paris Climate Agreement. The Secretary General may choose to place this **Climate Panel** under the UNCCC (United Nations Framework Convention on Climate Change) but that choice is up to him.

Figure 62 UN Agency that oversees the Paris Climate Accords.

4. The **Climate Panel** will be fully responsible for administering this game and for guaranteeing 100% transparency.
5. All entrants (country, state, company, school, tribe, family, or individual) will submit an application form stating the intention to enter the contest together with the agreement if they do not win, to donate 1% of their GNP (for countries), or gross income (for companies and individuals) as prize money for three consecutive years.
6. When an entrant has discovered a new source of electricity that is clean, efficient, and is capable of large-scale implementation, he will submit the design, proof of concept, and experimental results to the **Climate Panel.**
7. The **Climate Panel** will see that all entries are posted on a public website and that the website is regularly updated with progress reports from the entrants.
8. The game will continue until there are one or more winners as determined by the **Climate Panel**. Entrants can opt out at any time and from the date of that opt-out, if five years go by without a winner, they are freed from the obligatory three 1% payments.

9. The **Climate Panel** will review entries and determine the winner or winners.
   (a) If more than one winner the reward is shared.
   (b) The rewards for winners of the secondary prize (Rule #2 above) will be determined by the **Climate Panel**.
10. The **Climate Panel** will ensure the winning technology is made freely available to the entire world and in no case will any country, company, ideology or class be allowed to use the technology at the expense of the rest of the peoples of Earth.

The rewards

The country, state, company, university, individual, or other entity who wins, who produces the first actual new workable alternative, renewable, cheap, dependable, clean power source, is rewarded with three yearly payments each of 1% of the GDPs of every nation participating in the game and three yearly payments each of 1% of the yearly gross income of all other entrants.

The roll-out of a new clean source of electricity will offer enormous opportunities to all countries, both the rich and not so rich. Besides the huge numbers of new well-paying jobs, these opportunities will provide rich targets for public and private investment, and include unlimited chances for entrepreneurs to manufacture and provide and service the direct and ancillary items that will be needed to get the new source(s) of energy into global use.

Some countries will be better equipped to help others with the design and construction of their new energy systems, and such assistance will provide commercial opportunities. It is inevitable that an energy-miracle rollout in accordance with this proposal will go far towards decreasing the global gap between rich and poor.

But the greatest reward is that the basic technology behind the Energy Miracle is being made freely available to all.

The secrets

(1) You don't have to be a quantum whiz to win at this game because none of the other scientists and inventers of history's top 42 electrical inventions had any connection to quantum

mechanics. In fact, most of them totally rejected the ideas currently held by quantum mechanics. Think of that.

(2) Governments, corporations, and universities would be well advised to throw human and financial resources at the problem with the expectation, besides the benefits to the Earth and mankind, that if they win, they would enjoy 1000+ times return on their investment.

(3) Individuals and groups with innovative ideas could appeal to their governments, educational foundations, industrial and other investors to fund their research and development actions.

(4) Entrants needn't even have a university degree. Half the inventers on the Top 42 electrical inventions list[40] had little or no formal education or were self-taught in the sciences. Among others, William Gilbert, Francis Hauksbee, Stephen Gray, André-Marie Ampère, Michael Faraday, George Ohm, Hippolyte Pixii, and Benjamin Franklin were all self-taught. Thomas Davenport was a self-taught blacksmith. Faraday's father was also a blacksmith and he himself was a bookbinder. Jacob Schoelkopf was apprenticed in a tannery. Very few on the Top 42 list were university educated.

(5) Bill Gates gives us roughly 10 years to come up with a workable energy miracle, and 20 years more to get it into widespread use. This means any student, from middle school to college, will have enough time to set his sights on this goal, go after it, and achieve it.

(6) You needn't be a white Anglo-Saxon male to win despite the fact that all the inventors on the Top 42 list were of that category. The world has changed. The field has opened up. Black or brown, female or LGBGT, Chinese or Russian, Hindu or Muslim or South Sea Islander or Japanese or Native American....any and all can enter and have an equal chance to win.

(7) This program can be gotten off the ground very quickly and spur a lot of very smart minds to work at solving the energy challenge. What we're facing is an ENERGY crisis, and this is a direct address to the problem. The structure of the

[40]See Table 2.

program and its incentives will make it economically viable for investors. And in the end, it will benefit all with a low cost, reliable, clean source of energy.

(8) Seize this chance. Study the basics of energy as can be found in this book, figure out a new way to harness electrical energy, and let no power or persuasion deter you from your path.

PART II

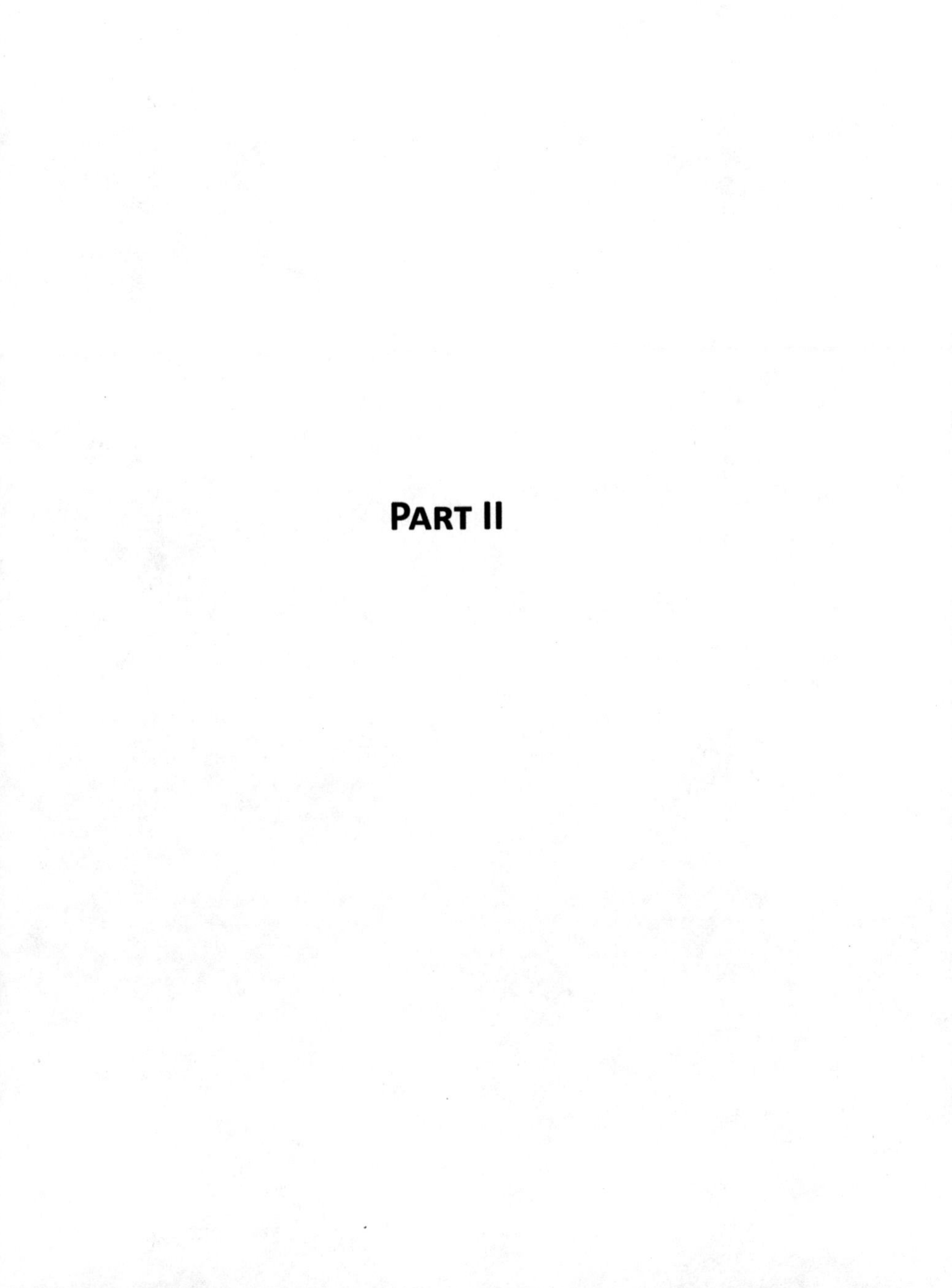

PREFACE TO PART II

The (Not-That-Technical) Technical Side of the Energy Miracles

In the second part of the book, we delve into some of the weeds of quantum mechanics that have been blocking the search for new sources of electrical energy. Starting with an evaluation of exaggerated claims made by quantum mechanics, we look, one by one, at a dozen or so of the quantum mechanical theories that have succeeded in tying the brain of science into a hangman's noose. For the first time in print, ungraspable concepts such as the uncertainty principle, entropy, spacetime, Schrödinger's Dead and Alive Cat, and the fourth dimension are elucidated in clear English without resort to the subterfuge of forbidding technical terms or mathematics.

This is followed by a look at successful methods instructors can employ to re-energize energy students, plus a recommended study program helpful to those in search of the Energy Miracles.

A dozen of the most promising energy sources are reviewed where the five keys to the Energy Miracles could be well employed, and the final section is a discourse on one of nature's most accessible examples of energy: lightning.

Chapter 8

Quantum Quagmire: Dead End for Energy Miracles

"I think I can safely say that nobody understands quantum mechanics."[41]

—Nobel Laureate Richard Feynman

Yes, we are going to peruse the subject of quantum mechanics. We have been left with no choice. When you power up your new computer only to find out that you are totally blocked from communicating with your friends, that half your photos are missing, and that Internet searches return mostly advertisements, you cannot avoid taking some interest in its workings. Same with energy science—there is no reason we have to stick ourselves with 19th-century energy sources. Earth cannot afford that luxury.

Quantum mechanics have been described as people who know more and more about less and less, until they know everything about nothing. In the long history of science, all advances have been based on somebody having found some truth. Quantum mechanics is the first science based on not having found it.

Folks generally avoid quantum mechanics on account of its complexity. Its proponents preach that the only way to understand it is to glorify its mystery or (even worse) to make it even **more** incomprehensible. Rather than go along with that, we are going to shed light on a dozen or so of its most popularized and infamous

[41]Richard Feynman, "Probability & Uncertainty: The Quantum Mechanical View of Nature," A lecture given on November 18, 1964, at Cornell University, U.S.

Energy Miracles: The Global Warming Backup Plan
H. B. Glushakow

ISBN 978-981-4968-18-8 (Hardcover), 978-1-003-28442-0 (eBook)
www.jennystanford.com

shadows as they pertain to the discovery of Energy Miracles. People sometimes become fearful of deep shadows but notice how quickly a shadow dissipates by the simple operation of shining light on it. We are not going to push mathematical formulas at you or theoretical physics. We are going to use common sense, logic, and observation. And when we are done, we hope any anxiety or unease the subject may have held over you will have lifted. But more importantly, we will show how these theories have been blocking the search for Energy Miracles.

Figure 63 Prof. Snodgrass commenting on quantum mechanics.

A Little Quantum Background

What does quantum mechanics consist of? What are its principles? Quantum—the name of the theory that energy comes in finite packages and is not infinitely divisible—was introduced in 1900 in an attempt to explain puzzling results in physics about how light was emitted and absorbed.

Experts describe the progress of the subject in three parts: the initial stage of quantum physics (between 1900 and 1925); the second stage, which ushered in what is called quantum mechanics (1925–27); and the post-1927 period in which quantum electrodynamics (QED) evolved. For simplicity, in this book we lump all three together as quantum mechanics since they are all based on the same flawed foundation.

There is a peculiarity of quantum mechanics that exists in few places in science: Its main advocates, experts, gurus, and professors, even after 120 years, cannot agree on the basic concepts or implications of the subject. For example, despite thousands of papers that refer to, discuss, and criticize the Copenhagen Interpretation of quantum mechanics, there is no place that you can go for a concise, agreed upon, statement that defines what the Copenhagen Interpretation even means.[42] Up to the 20th century, physics was the science of studying and understanding nature and the things around us so that we could better control them and make better informed predictions. Quantum mechanics redefined all that at the beginning of the 20th century when physics changed its role to the promulgator of mathematical formulas to explain theoretical presumptions that were not supported by empirical[43] data. In classical physics, one would propose a theory and then see whether empirical data would support it. The quantum mechanical way is to dream up a theory and "prove" it by creating a mathematic formula. The joke is that mathematics is just an accessory, and a clever mathematician can **always** manufacture a formula to prove a theory.

Einstein's General Theory of Relativity could be called the mother of quantum mechanics. But a sad mother she must be. Today's quantum mechanics tell you that Einstein's relativity is not a quantum mechanical theory at all, and that it is not the way the universe should be understood. The "new, correct" way is something called the "Standard Model of Particle Physics," which describes the behavior of matter solely in terms of its elementary constituents, those constituents being invented particles with invented characteristics for which there is absolutely no empirical support—small things like neutrinos and quarks that you would not find under a microscope. So, Einstein invented his relativity and the newbies invented their imaginative particles and both sides insist the two are incompatible.

---

[42]The Copenhagen Interpretation is an expression of the meaning of quantum mechanics devised between 1925 and 1927 by Neils Bohr and Werner Heisenberg, but there are many interpretations of that interpretation. Though widely cited, and commonly taught, its exact meaning and implications remain variable and obscure.

[43]Empirical: based on, concerned with, or verifiable by observation or experience rather than theory or pure logic.

By 2011, quantum mechanics had reached its 100th year; and between 2011 and 2013, organizers of four advanced quantum conferences (one in Baltimore, MD, U.S.; one in Austria; and two in Germany) surveyed 175 of their attendees to ascertain their views and opinions on the basics of the science. They used the exact same survey questions in each of the four conferences. It might be expected that a century was enough time for quantum adherents to forge some agreement on the meaning and implications of their discipline, but the survey results showed the reverse. The basic tenets of quantum mechanics were not only unresolved but still being very hotly contested. Not only that, the most recent of those surveys noted significantly more controversy over fundamental issues than was evidenced in the earlier ones. Are the quantum mechanics facing a nuclear meltdown from within?

Quantum mechanics (QM, also known as quantum physics or quantum theory, or quantum electrodynamics or QED) was originally conceived as a branch of physics dealing with physical phenomena at microscopic scales. That was somewhat of a misnomer on two counts. First, since the particles involved were way smaller than could be seen through any microscope, the theories involved were all based on, well, nothing very substantial. And second, the subject quickly swept over the dam of microscopic scales and began claiming it explained not only those things at the microscopic level, but the character and behavior of **all** energy and matter in nature. That usually goes along with: "So criticize it at your peril."

Figure 64 A quantum mechanic calculating how many angels can fit on the head of a pin.

There is sometimes confusion between the terms quantum mechanics (the subject) and quantum mechanics (the adherents and disciples). Interchanging the actual subject with often-grimy fellows who claim to be able to fix used auto engines[44] is an easy mistake to make, and I plead guilty to it. So, when you see the term quantum mechanics being used in this book, realize it could refer to either the subject or to its card-carrying devotees. Figure 64 is one of the world's more popular portraits of a quantum mechanic. If you look closely, you will see the artist caught him in the act of dreaming up a theory for how best to calculate the number of angels that can fit on the head of a pin.

In this chapter, we shall first take up the claims the quantum mechanics make about themselves. The rest of the chapter is devoted to some of the more popular aspects of quantum mechanics. A few of the subjects chosen have received some notoriety (such as Schrödinger's Dead and Alive Cat and Heisenberg's uncertainty principle), but all of the points taken up here were chosen on account of their roles in obstructing the discovery of 21st-century Energy Miracles.

An Ode to the Quantum

Quantum queer, to make it clear, its point seems indefensible.

We'd prefer to shed some light and make things comprehensible.

In days gone by, the clean straight look was valued quite a treasure

Now to observe is disallowed; much less to try and measure.

The Quantums say they're not short-term, but bound to last forever.

To even try to understand is wasted time endeavor.

So, without shame they raise their cups, in dutiful libation.

To toast the end of science and of real observation.

Although these Quantums do their best to keep us all confused,

What's true for you is true, no matter how much force they use.

The Quantums think all nature's truths are hid behind a curtain.

Yet knowledge, true, is only that of which you've become certain.

Quantum Claims and Exaggerations

"The more success the quantum theory has, the sillier it looks."

—Albert Einstein[45]

[44]No disrespect intended for real mechanics who fix real cars.

[45]Letter to Heinrich Zangger, May 20, 1912: CPEA, Vol. 5 Doc. 398.

The word quantum permeates our world. In popular usage, we see myriad uses of quantum this and quantum that, of which 99% are meaningless buzzwords trying to impress, dazzle, or persuade money from customers or educational and government funding sources. We are told quantum mechanics is the science that has shaped our world. We now have quantum beer and quantum jazz and quantum apocalypses. The word quantum is applied to almost every aspect of science, to psychology and neuroscience, to aviation, politics, entertainment, sound, and even fishing.

Figure 65 "Let's just say there's been some exaggerating going on around here..." (drawing by M. J. Coyle).

Serious people tell us that we are indebted to quantum mechanics for just about every underpinning of our culture and economy. The most brazen claim is that quantum mechanics created the Information Age.

We were forced to look into quantum mechanics when we found its tenets interfering with the search for Energy Miracles. It is neither seditious nor irreverent to question quantum claims about energy, when we find so many of its other major claims to be misrepresentations or outright fabrications.

Quantum mechanics insists we are indebted to it for many reasons. Seven of its most important claims appear in Table 3. Amazingly, none of these alleged accomplishments had much if anything to do with quantum mechanics.

Table 3 Quantum mechanics top brags

#	Quantum Claimed Invention	Alleged Importance	Actual Origin of the Invention (None Are Based on Quantum Mechanics)
1	**Semiconductors**	They are the basis of the information age and come from Quantum Mechanics.	Michael Faraday discovered the semiconductor effect in 1833. Karl Ferdinand Braun established the basic phenomena of conduction and rectification in semi-conductors in 1874.
2	**TVs, Smart Phones, LEDs, MRI Scanners, Computers**	These items contain transistors and transistors come from Quantum Mechanics.	J. E. Lilienfeld patented the first FET (field-effect transistor) in 1925, based on the positive and negative behavior of energy as described in this book. Lilienfeld's patent includes no slightest reference to any quantum mechanical theory. See next section.
3	**Quantum Computers**	Because they are so much faster & more powerful than regular computers.	There are no quantum computers. After 20 years of research and billions of dollars of investment, there are no useable quantum computers. None.
4	**Quantum Encryption**	Because code breaking is so important in our society.	There are no quantum encryption or de-encryption programs. 20 years of Defense Dept. funding has not produced a single product. None.
5	**GPS**	Because it relies on atomic clocks.	Lord Kelvin suggested the use of atomic transitions to measure time as early as 1879, long before quantum mechanics.
6	**Periodic Table of Elements**	The Periodic Table is fully explained by Quantum Mechanics.	The Periodic Table was formulated 150 years ago, long before quantum mechanics, and recent studies show that quantum mechanics cannot account for the properties of 20% of the elements.
7	**Gravitational Waves Move at the "Speed of Light"**	Gravity is explained by the "curvature of Spacetime."	Isaac Newton discovered gravitational waves in 1686. Nowhere has quantum mechanics shown how the geometric abstraction called "curved space-time" is able to create a gravitational wave that propagates at the speed of light.

It is not as if there are lots of other things we can point to for which we can give thanks to quantum mechanics for providing to us. The preceding seven things are the things they claim they have provided. And they are all pretty much a bunch of smoke and mirrors.

Quantum's most audacious claim

"I will never forget the day I first met the great Lobachevski.[46]
In one word he told me the secret of success in mathematics:
Plagiarize. Plagiarize!
Let no one else's work evade your eyes.
Remember why the good Lord made your eyes.
So don't shade your eyes
But plagiarize, plagiarize, plagiarize.
Only be sure always to call it, please, "research."
—Tom Lehrer, Harvard and MIT mathematician and satirist

Oh, if that were only satire. We introduce this section with the quantum mechanic description of how the Information Age was born:

"Once upon a time, in 1947, a wonderful Quantum Mechanic named William Shockley used the theories and mathematics of quantum mechanics to invent the transistor, which is the basis of all electronics and thus the basis of the Information Age. In fact, Shockley started Silicon Valley all by himself."

William Shockley was a brilliant theorist with 80 patents to his name. Some of his colleagues and a few of his students praised him highly and he won many awards, but as we shall see, he had very little to do with the birth of the Information Age.

The electrical signal bringing sound to your smartphone is far too weak to be heard by the human ear. So, its intensity must be boosted. This was originally accomplished in radios and televisions with vacuum tubes, cousins of the common light bulb. But these were too bulky, consumed too much power, created too much heat, and had too short an operating life. The transistor was the solution to all

---

[46]Russian mathematician whose "hyperbolic geometry" was used in the creation of quantum spacetime.

of those issues and, once developed, became the key component of all electronics. And although the importance of its initial discovery and further development cannot be overstated, none of the claims of quantum mechanics about its source are true.

In the first place, Shockley did not invent it. The transistor was actually invented by Julius Edgar Lilienfeld in 1925. Lilienfeld was another brilliant engineer with over 60 patents to his credit. He was a hands-on guy who knew electrical theory but challenged it through his experiments. His three transistor patents are based on his observation of the positive and negative behavior of energy as described in this book and include no slightest reference to any quantum mechanical theory or mathematics.[47] How the myth came about that the transistor and Internet revolution were children of quantum mechanics makes for an interesting story.

Lilienfeld discovered, designed, and patented the transistor in 1925. He also built working models. The semiconductor physicist H. E. Stockman recalled, "He (Lilienfeld) created his non-tube device around 1923, with one foot in Canada and the other in the USA, and the date of his Canadian patent application was October 1925. Later American patents followed. Lilienfeld demonstrated his remarkable tubeless radio receiver on many occasions."[48]

Lilienfeld's "field effect" invention can be simply described: a device able to amplify an electrical current in solid material by applying an external electrical field to it. His device indeed worked, but not well enough to be developed commercially without a high-purity semiconductor (a material that neither conducts nor insulates electric current, but lets it pass variably). In 1925, high-purity semiconductor materials were not yet available. But by the mid-1940s, many teams were on the track of developing such material. The aforementioned William Shockley was leading the team at Bell Labs. The most important members of his team were John Bardeen and Walter Brattain, who were to make the breakthrough that would win for them the Nobel Prize in Physics.

It is uncertain when Shockley first saw Lilienfeld's patents, but it is certain he was familiar with them before his team's big 1947 breakthrough. He could have had them as early as 1936 before

[47]Lilienfeld's patents are posted on www.energymiracles.net.

[48]See Bell Labs Memorial: "Who really invented the transistor?" https://www.beatriceco.com/bti/porticus/bell/belllabs_transistor1.html

World War II when he first came to Bell Labs and was tasked with replacing the vacuum tube with "a more stable, more solid and cheaper device."[49] It could have been in 1945, after the war when he put together the team of Bardeen and Brattain. But it was certainly no later than the year before he submitted his patent applications because the legal files that led to the original Bardeen and Brattain patent included an affidavit submitted by J. B. Johnson, a well-known Bell Labs physicist, testifying that Shockley had previously attempted (unsuccessfully) to build the exact device described in the Lilienfeld patent.

What we do know is that despite the full weight of Bell Labs behind him, Shockley had been stymied for nearly 10 years in his effort to use the theories of quantum mechanics to make Lilienfeld's idea work. Though he was the nominal director of the project, by 1947 Shockley had not been personally involved with the day-to-day workings of his team for almost 2 years. Possibly discouraged by his earlier failures, he had gone on to other projects and left the work of solving the semiconductor problem to Bardeen and Brattain. Their breakthrough was preceded by their decision to discard Shockley's quantum theories and replace them with some basic chemistry, some classical ideas on electricity, and some hands-on experimentation.[50] Once they did that, and without Shockley's further participation, using little more than a tiny piece of the element germanium, a thin plastic wedge, a paper clip, and a shiny strip of gold foil, they came upon a different and successful way to make such an amplifier. Their voltage-controlled device could boost an electrical signal by 100 times.[51]

After confirming their results, they informed Shockley of their discovery, and an exhibition was scheduled for Bell Labs executives on December 23, 1947. But Shockley was not happy. "I experienced frustration that my personal efforts, started more than eight years

---

[49]These instructions could have been taken directly off Lilienfeld's patent, which made those EXACT claims: to replace the vacuum tube, to be simpler and more substantial, and inexpensive.

[50]See, for example, Bardeen's notebook entries of November 23, 1947.

[51]We know this to be the case because Bardeen and Brattain noted it in their laboratory logbooks. See Joel N. Shurkin, *Broken Genius: The Rise and Fall of William Shockley*, Macmillan, New York, 2006, and Arns (previously cited). For more background, see Michael Riordan and Lillian Hoddeson, *Crystal Fire: The Invention of the Transistor and the Birth of the Information Age*, W. W. Norton & Company, New York, 1998.

before, had not resulted in a significant inventive contribution of my own." That knowledge was not to stop him, in early January, from calling Bardeen and Brattain into his office separately, to announce his intention to be the sole person on the transistor's patents. Bardeen was left speechless and just stormed out of the room. Brattain snapped back: "Oh hell, Shockley, there's enough glory in this for everybody!"

Bardeen and Brattain's original point-contact FET transistor inspired a flood of activity aimed at creating new devices that could control the flow of electrical charge in solids. Just a few weeks later, in early 1948, building upon that discovery, Shockley himself invented a slightly modified transistor he called a BJT (bipolar junction transistor). Unlike the FET, Shockley's BJT was controlled by current. Shockley pressed the lawyers for a broad patent based on his claimed original idea of how a field effect could influence the current in a semiconductor, but they refused, due to the precedence of the Lilienfeld patents. In the end, Bell's attorneys filed four patents. Two were on the initial work Bardeen and Brattain had done exploring Lilienfeld's field-effect transistor. One for the actual device Bardeen and Brattain created (Bell's original field-effect transistor or FET). And the final one for the modified transistor Shockley finalized a few months later (the BJT). All four of these patent applications were filed in the summer of 1948, just before they were announced publicly.

The U.S. Patent Office quickly rejected the first two patents, finding they infringed on/copied Lilienfeld's work. The second two—the one for the FET point-contact transistor and the other for Shockley's BJT—were deemed acceptable by the Patent Office, but only with conditions. As for the application for Bardeen and Brattain's original FET, all but 19 of the 69 claims made in that application were denied. Thirty-six claims involving the basic ideas of transistor action were disallowed outright mostly on the basis of the earlier Lilienfeld patents. As for the application for Shockley's BJT transistor, of the 62 claims he made, only 34 were granted by the Patent Office, the most frequent reason given for the rejections again being Lilienfeld's earlier drawings. A total of 65 claims were either abandoned or disallowed in these two patents, with Lilienfeld's prior work the most cited reason. Legal papers from the Bell Labs patent show that Shockley had attempted to build operational versions directly from Lilienfeld's patents. What is more, he never referenced Lilienfeld's work either in his own patent application or in any of his later research papers, historical articles, or books.[52]

Bardeen set the record straight in 1988 by acknowledging that in the 1920s before the quantum theory of solids was proposed, "Lilienfeld nevertheless had the basic concept of controlling the flow of current in a semiconductor to make an amplifying device."[53]

A senior quantum mechanic at the time, attempting to defend Shockley and believing he was launching a valid criticism at Bardeen, later wrote: "His (Bardeen's) thoughts were blocked by his classical electrical-engineering-oriented thinking during that period."[54] Yet it was that exact classical electrical-engineering orientation that enabled Bardeen (not Shockley) to make the breakthrough that led to the FET transistor.[55]

Shockley was not poorly treated. Bell Labs included him in all statements and photos concerning the FET discovery, and he was allowed to share the Nobel Prize in Physics for it together with Bardeen and Brattain. But he refused to be mollified and remained furious that his name was not on the patent for the original FET that would replace the vacuum tube. More pertinent to our theme here is how he attempted to control the narrative about it and elevate his role in its discovery. There is a joke about quantum mechanics that goes: "Now that they know something works, they've got to figure out how to prove it was actually predicted by quantum mechanics." And that is exactly what Shockley did. He first prepared a paper for *Physical Review Letters*, but the editors rejected it, admonishing him that his explanation of the quantum theory behind his device was insufficiently rigorous.[56] Then, in 1950, he published his 545-page *Electrons and Holes in Semiconductors*,[57] in which he popularized imaginary "electron holes" and "charge carriers" to prove that the

[52]R. G. Arns, The other transistor: early history of the metal-oxide-semiconductor field-effect transistor, *Engineering Science and Education Journal*, **7**, 5, 1998, 233–240.
[53]Bardeen's letter to William Sweet, Associate Editor of *Physics Today*, dated March 9, 1988.
[54]P. K. Bondyopadhyay, In the beginning, *Proceedings of the IEEE*, **86**, 1, 1988.
[55]In Chapter 4 of *The Innovators* (Simon and Schuster, 2014), Walter Isaacson gives a detailed and readable account of the development of the transistor.
[56]E. Braun and S. MacDonald, *Revolution in Miniature: The History and Impact of Semiconductor Electronics*, Cambridge University Press, London, 1978.
[57]Published by D. Van Nostrand Co., Inc. Princeton, NJ, 1950.

transistor was actually fundamentally derived from his work using quantum mechanics theory. That book was taught to almost every electrical engineering class in the English-speaking world. **That is the source of quantum mechanics claims that quantum theory alone is responsible for the Information Age.** But as shown earlier, it is a totally unwarranted boast.

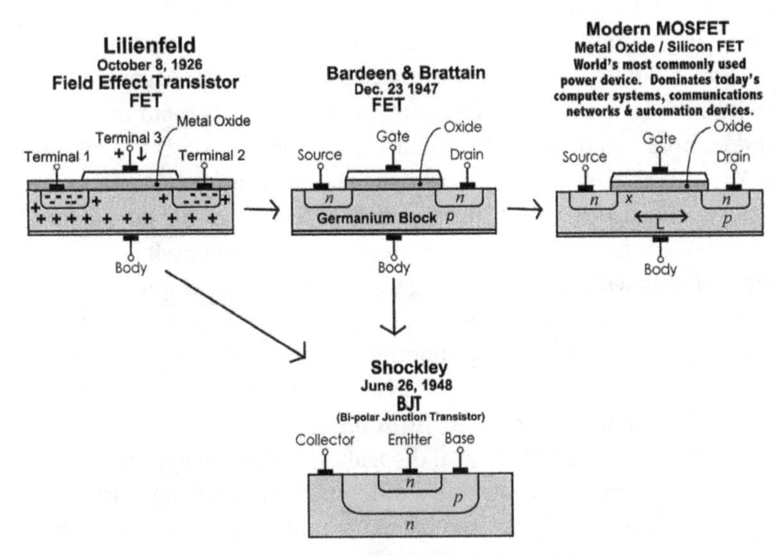

Figure 66 Transistor timeline. Today, almost 100 years after Julius Lilienfeld invented the FET (field-effect transistor), MOSFET transistors based on his design dominate semiconductor electronics.

Shockley did not invent the transistor. He copied Lilienfeld's 20-year-old patent and personally made little or no contribution to the Bardeen/Brattain breakthrough. This is not to say that he made no contribution at all. Shockley's BJT was more practical than Bardeen's FET back in the 1950s and became the main active device in first-generation mainframe and minicomputers. But the BJT's functionality was largely replaced half a century ago, and today's computer systems, communications networks, and automation devices are almost all based on descendants of Bardeen's FET. Due to their many technical advantages, these MOSFETs are the most commonly used power devices in the world and dominate integrated circuit technology. Shockley's less flexible BJTs, though still in use today, are limited to switches or amplifiers in applications such as low-power LEDs, temperature sensors, and some other specialized cases. So, yes, Shockley was a contributor to the Information Age, but that is a long way from saying he was its prime mover.[58] (See Fig. 66 timeline of FET diagrams.)

The Franklin Institute politely summarizes this matter as follows: *"The exact nature of Shockley's contributions to the development of the transistor remains a subject of controversy, as does the question of how much (if any) credit he should be given for its invention."*[59]

As for single-handedly starting Silicon Valley, that turns out to be a backhanded compliment. After threatening his team members and company management at Bell Labs, Shockley left the company in 1955 and set out to make his fortune by starting his own company near Stanford University in California. Attempts at hiring a bunch of engineers he had worked with at Bell Labs failed, when they all refused. But with the help of the Stanford Dean of Engineering, he hired nine competent engineers with PhDs. Although he was in the unique position of providing something of great value to a civilization that was craving it, the venture was a failure almost from the outset. It took less than a year for eight of his top nine engineers to bolt from under his erratic technical theories and management practices. Shockley's company was losing a million dollars annually and a few years after the "bolt," with the company under water, Shockley was forced to sell it. As for the eight who bolted, they all went on to great successes in Silicon Valley and made billions of dollars. Two of them founded the Intel Corporation. This was not the first time Shockley had blown away good talent. After showing his true colors at Bell Labs, his star team members (Bardeen and Brattain) both quit his team and refused to ever work for him again.

That Shockley would be capable of stealing someone else's patented design and then seek to falsely legitimize it with pseudo-scientific theories may be better appreciated by observing another arena where he did a similar thing. In the interest of full disclosure, I must admit my tracks crossed those of Dr. Shockley back in the 1960s when I was busy doing things like registering voters in Mississippi, teaching black children in Virginia (when white supremacist politicians chose to close the schools rather

---

[58]And, of course, we must remember it was Julius Lilienfeld who made the original discovery.

[59]https://www.fi.edu/case-files/bardeen-and-brattain

than integrate them), being hauled away by police at the 1963 March on Washington, and arrested for nonviolently protesting the segregation of lunch counters at the New York World's Fair. At that same time, Dr. Shockley was embarked on a personal mission to persuade the world that black people, Chinese, and the poor of any color were genetically inferior to the white. He called the *concept of equal rights* America's "national egalitarian lie," and his pseudo-science became front-page news. Naming it "raceology, the scientific analysis of racial differences," he became the number one voice in America railing against social justice. He received more bandwidth than the KKK with pronouncements such as: "*The major cause for American Negroes intellectual and social deficits is hereditary and racially genetic in origin and thus not remedial... Nature has color-coded groups of individuals so that statistically reliable predictions of their adaptability to intellectually rewarding and effective lives can easily be made and profitably used by the pragmatic man-in-the street.*"

Shockley's words are enshrined in his 300-page book *Shockley on Eugenics and Race: The Application of Science to the Solution of Human Problems* in which he uses quantum mechanical theories and psychology to try and convince us of the superiority of the white race. Invoking Einstein, Heisenberg, and Feynman, Shockley quotes the Lorentz force law and special relativity to make his case that it is necessary to suppress two-thirds of the Earth's population.

His book shows Shockley's love for Einstein "thought experiments" as covered later in this book. One thought experiment described by Shockley showed how the world might reduce the impact of "poor genes" in the society by offering large bonuses ($30,000) to any person of color who would agree to be sterilized. Another even darker one envisioned the mandatory insertion of intra-uterine contraceptive devices in all women of color, each to be given an accompanying certificate allowing her device to be removed just long enough for her to have one child. The problem with his "thought experiments," besides their inhumanity, is that they are founded on baseless theories. No study has ever found any slightest link between IQ and genes. The reason is that there is no such link. Just as he did with his quantum mechanics, Shockley just made the data up.

Barred from lecturing at universities, Shockley nevertheless followed his quantum mechanical philosophy to his grave. He died, alone and generally despised, except by his wife. How quantum mechanics impeded him can be seen in his answer to a question from an interview included in his book on race.

Question: *"It may be helpful for us to know something about the tenor of your personal relationships with blacks. It could give us some insight into your motive."*

Shockley: *"I basically haven't had much personal contact with blacks."*

If you never make the effort to look at something and communicate with it, you will never understand it. Quantum mechanics, including Dr. Shockley, invent theories concerning subjects about which they have done little or no personal observing or communicating. Much more on that in the next chapter.

Quantum Mechanics Contribution in the Fight Against Global Warming

"You don't have to be an ichthyologist to know when a fish stinks."

—A. Unzicker

If not any of those seven aforementioned exaggerated claims, then what? What do the quantum theories bring to the table?

If nothing else proves the relevancy (or lack thereof) of quantum mechanics, it is the contribution it has made in confronting and mitigating the biggest challenge of our times: global warming.

In one word, what quantum mechanics has contributed to that work in the past 30 years is:

Nothing.

Science before the Quantum

Before quantum mechanics, the discoveries of Sir Isaac Newton (1642–1747) permeated every aspect of western thought for over 200 years—not just in science, but arts, literature, philosophy, and religion. Newton introduced clarity and intelligibility to an era of confusion and uncertainty. Before Newton, wars and plagues were thought to come from supernatural causes. Before Newton, only kings and the clergy were thought to have any understanding of causes—due to their special connections to God's will. The rest of humanity was steeped in a soup that was unknowable, unpredictable, and widely fearful. Newton changed all that in 1665 when, taking advantage of the opportunity of uninterrupted study at his mother's farm during the Great Plague that killed over 20% of London's inhabitants, he made many fundamental discoveries in physics, astronomy, mathematics, and optics. He gave humankind its first picture of a world that was unified and comprehensible. For the first time, humanity was given a thoroughly believable introduction to the concept of cause and effect. Phenomena and conditions could now be explained in terms of predictable forces. Anyone who could read could understand the world. The quantum revolution changed all that.

The question is repeated: "Why has there been no major advance in the subject of electricity for almost 100 years?" And here we answer it.

Figure 67 A quantum mechanic taking the dead-end path.

A century ago, science made a choice. It abandoned a line of research and discovery in electricity that had succeeded in changing the course of history and modernizing much of the world. Abandoning logic and scientific methodology, scientists began to concoct and follow new and mysterious theories that no one could understand, were unproveable, and yet were somehow intriguing enough to stick one's attention.

So here we delve into those roads, both the ones taken and the ones abandoned, in an attempt to provide some orientation and understanding, if not enlightenment, about what happened.

To repeat a point made earlier: A time-honored practice for reversing non-optimum situations and restoring positive trends is to discover **what changed**. By nullifying a disadvantageous change in course, or restoring a successful one, one can usually get back on track.

In the case of the abrupt cessation of advancement in the field of electrical energy, if we could put ourselves back on the path that was obtaining results from the point where we fell off, we should immediately be able to resume harvesting the kinds of advances that were occurring before we took that unfortunate fork in the road.

And the world really does need an Energy Miracle.

Quantum Mechanics: A Mystery in a Monastery

Whenever someone sticks a gun to my head and tells me he knows that what he is about to say would not make sense to me, that he knows it is not provable in any way, but that I had better believe it nonetheless "or else," my reality meter starts to twitch and I am reminded of the story of the monastery on the hill. This was not your normal garden-variety monastery, for it held the answer to the mystery of all mysteries: *the truth that explains all.* It was heavily fortified against intruders and for centuries had remained resistive to all attacks. Yet in legends, it had been foretold that on a certain day that monastery was going to fall.

On the day predicted for that catastrophe, a local peasant, who had been dutifully paying his tithes to the monastery his entire life and held it in the greatest awe, decided to climb up the high hill on which that monastery stood, and do the unthinkable: enter and discover for himself the answer to the mystery of all mysteries. He approached the front door without challenge (which was strange) and found the door open, which was stranger still. Stepping through the huge portal, he still encountered no interference. The few soldiers present were crying in the corner because the monastery was supposed to fall that day. Still unchallenged, he walked through room after room, humbled by the fine tapestries, stained glass windows, and splendid oil paintings. Making his way deeper and deeper into the unguarded monastery, he finally arrived at the inner sanctum. Its 15-foot solid teak door was decorated with inlays of gold, a mysterious "Q" crusted with precious jewels at its center. Fearful but determined, he laid his hand upon the giant brass ring and watched the door open smoothly to his touch. Inside he found a very quiet room, with no windows but with floors, walls, and ceiling all paneled with hand-carved exotic woods of many kinds. By the light of the candles in bright brass candelabras, he saw, at the opposite side of the room, a small alcove behind black curtains.

And he knew, finally, that he had arrived at the destination he sought. Approaching the inner chamber, he pulled back the curtains to find the long-hoarded secret of the mystery of all mysteries waiting for him there. He looked. The room was bare. There was no answer; there was no mystery. The whole thing was a fraud.

But there was truth to the legend. On that day, the scam was finally revealed. And on that day, the monastery fell.

Foundations of Quantum Mechanics: A Grim Fairy Tale

> *"Don't keep saying to yourself if you can possibly avoid it: 'How can quantum mechanics be like that?' because you'll get down the drain, you'll get down into a dark blind alley from which nobody has yet escaped. Nobody knows how it can be like that."*
>
> —Nobel Laureate Richard Feynman

> *"If you can't explain it to a six-year-old, you don't understand it yourself."*
>
> —Nobel Laureate Albert Einstein

In every book on the subject of quantum mechanics and in every lecture series, somewhere in prominence will be found these three assertions:

1. Quantum theories are "counter-intuitive" (meaning they defy common sense);

Figure 68 The weeds of quantum mechanics.

2. Quantum theories are "really" how things are (whether you believe it or not);
3. The study of quantum theories will "expand your view of reality."[60]

The first is most assuredly true: It spits in the eye of common sense. Since the second cannot be inspected, much less proven, it is at best a very uncertain opinion. As for the third, nothing could be further from the truth.

Quantum mechanics has long been introduced as "a complete theory of matter and energy based on the discovery that on the subatomic scale, energy comes in small discrete amounts." Remember, we are seeking to discover why the subject of electrical engineering ceased making forward progress in the mid-1930s. And we have found here that quantum mechanics, "a theory of matter and energy," gained popularity at precisely that moment in time. But what kind of energy theory? How do the quantum mechanics even define energy? One of the most important proponents of quantum mechanics, Nobel Laureate Richard Feynman, was honest enough to admit, "It is important to realize that in physics today, we have no knowledge of what energy is."[61]

In the remainder of this chapter, we will look at what quantum mechanics has to say about matter and energy and some of its major philosophical and scientific perplexities. It would be impossible to address the thousands of theories, formulas, and products that wear

[60]Frank Wilczek, *The Lightness of Being*, Perseus Books Group, 2008, p.6.
[61]R. Feynman, *The Feynman Series on Physics*, Addison-Wesley, 1988, 4–2.

the quantum mantle. Impossible and needless. What follows are a dozen or so that are generally agreed to be among the most basic and important, as well as impacting the quest for Energy Miracles. (Do not be scared off by the titles; most of them are easy reads.)

1. Quantum energy and Planck's constant
2. Major crack in the quantum mechanics foundation
3. The quantum end of the world: entropy
4. Science, energy, and Albert Einstein's house of cards
5. The Aether
6. The double-slit experiment
7. Heisenberg's uncertainty principle and its opposite
8. Schrödinger's Dead and Alive Cat (including a note from the cat)
9. Standard model of particle physics
10. TOE (Theory of Everything) or TON (Theory of Nothing)?
11. Spacetime and the fourth dimension
12. Origin of quantum mechanics: the uncertain zero
13. Quantum favorite sport: shoot the observer
14. Real experiments versus thought experiments
15. The observer: lost somewhere within the three universes
16. Insanity and the quest for an Energy Miracle

Quantum Energy and Planck's Constant

> *"Science investigates...Science gives man knowledge, which is power... Science deals mainly with facts and keeps us from sinking into the valley of crippling irrationalism and paralyzing obscurantism."*
>
> —Dr. Martin Luther King, Jr., August 30, 1959

The term "quantum" comes from the Latin word that denotes the amount of something. Specifically: "how much?" In physics, it means the smallest individual-separate-distinct piece of energy. Max Planck originated its use in science in 1900 when he used it in a presentation to the German Physical Society. Research into the effects of heat upon certain objects led Planck to the conclusion that radiation consisted of individual particles of energy. He called this particle a quantum.

If he had left it at that ("an individual particle of energy"), all would have been well. But he did not.

To understand Planck first requires an introduction to a ***black body***.[62] In physics, a black body is a theoretical physical body that has two characteristics. The first is that it should absorb every bit of radiation that hits it—any type from any direction or source. It should absorb it and then ensure none of it is let out. This is a highly theoretical imagining because no such "black body" exists in nature, nor has one ever been constructed. Not even close.

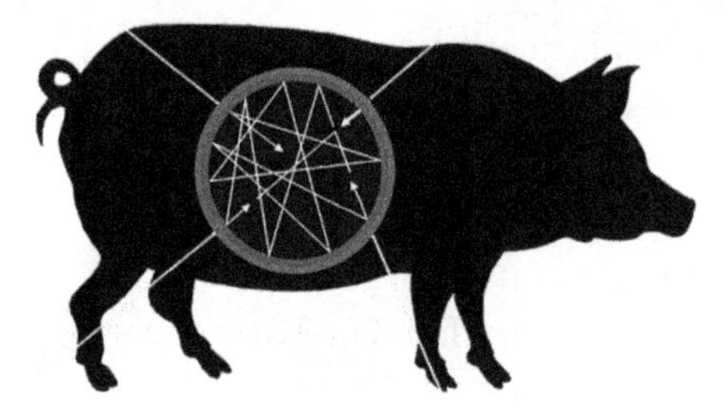

Figure 69 Max Planck's black body.

Planck postulated a second characteristic of a black body: If you cut a tiny hole in it, without letting out any of the radiation you had already trapped inside it, it should emit radiation of an intensity and range of wavelengths proportional to its heat. Only it did not. For one thing, to be seen, light must be reflected off some object, and there is nothing reflectable in a black body. So, it is not clear what exactly those black body experiments were measuring. More importantly, Planck was sure the answer lay in the effect of some other source of energy from the electromagnetic spectrum. He was right about that, but he could never prove it.

Figure 70 Hot piggy emitting light.

[62]Not Beyoncé's. This is a real scientific term.

Instead, he resorted to looking for a mathematical solution that would connect the **heat** of his theoretical black body with the intensity and frequency of the radiation it would theoretically emit. Try as he might, his equations would not fit the data of actual experiments; he just could not get them to match. But as any good quantum mechanic now knows, there is nothing that cannot be solved with a tad of imaginative mathematical fiddling. Using Ludwig Boltzmann's statistical methods and what he later called "a fortuitous guess," he devised a constant to be inserted into his formulas to make them work *under certain circumstances*. That constant, very small, unexplained, and unexplainable (approximately 6.62607×10^{-34} J-s) was later named "Planck's constant." He toyed with the problem of what significance, if any, might be given to his constant, but concluded that at best it was only a mathematical contrivance, a "fudge factor," the only use of which was in explaining the process of matter absorbing or emitting light under certain conditions. He strongly asserted that under no circumstances should it be taken as a description of the composition or character of light itself.

Planck's work around 1900 has sometimes been accused of being a sharp departure from classical electrodynamics, but the "quantum leap" attributed to Planck did not really exist until Albert Einstein gave it his novel interpretation. It was Einstein, in 1905, who took Planck's mathematical "fix" and turned it into a monster by declaring that light was definitely and incontrovertibly made up of discrete quanta the size of Planck's constant.

Before quantum mechanics, visible light had always been considered to be one very small section of the continuous range of energy wavelengths shown in Fig. 71. You could communicate with FM waves, cook dinner with microwaves, and see with light waves. But after Einstein, it had suddenly become "discrete quanta" the size of, or some multiple of, Planck's mathematical constant. There was no longer to be a continuous range of wavelengths of light; suddenly light would have to jump from one size to the next, much as in the underwear section of second-hand shops.

This had several immediate results, but the most important one was the invalidation of the greatest electrical discovery in the history of the world: Maxwell's theory of electrodynamics.

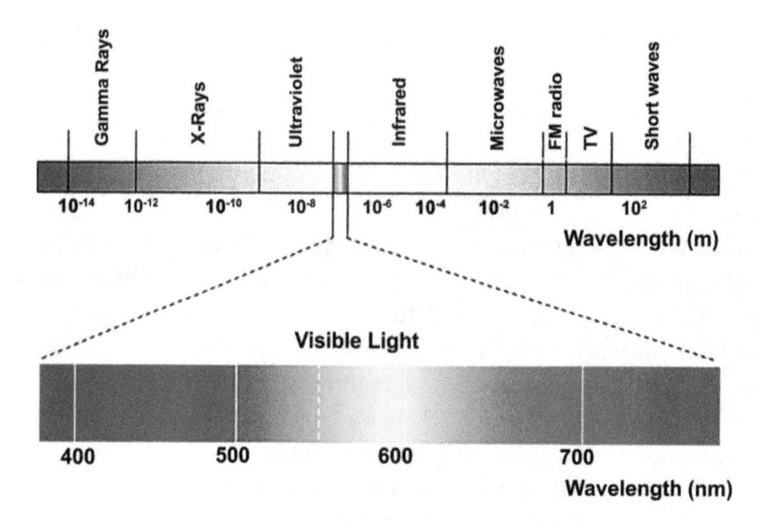

Figure 71 Range of continuous wavelengths of electromagnetic radiation.

Planck, an honest man to the end, panicked when he saw the use to which his work was being put and said prophetically that if Einstein's interpretation of his (Planck's) work prevailed, "The theory of light would be thrown back not by decades, but by centuries..."[63]

How precognizant he was!

Major Crack in the Quantum Mechanics Foundation

According to quantum mechanics, a *quantum* is the minimum amount of something that can be involved in the action of energy. For instance, a photon is supposed to be a single quantum of light, a single-point charge, the energy of which exists at a certain specific value according to Planck's constant. Same holds true for an electron.

Quantum mechanics distinguishes itself from the classical mechanics by claiming it alone works at very small scales. So, it is ironic that quantum mechanics has no solution to the problem of determining the electrical field of a single particle. In fact, this may be one of its most embarrassing and close-held secrets.

[63]Max Planck, *Scientific Autobiography and Other Papers*, Philosophical Library, 1968.

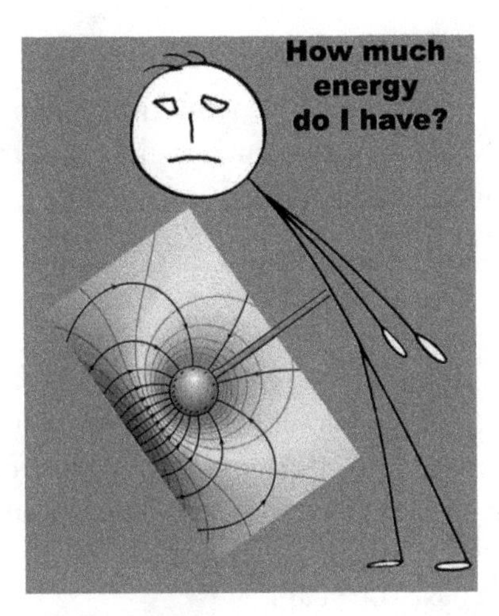

Figure 72 How much energy does a single-point charge have?

Classic electromagnetism tells us that the amount of charge or energy between two points increases as they come closer together or decreases as they move farther apart. (It is expressed as a formula that can be found at the end of this book in the Mathematics Postscript.) This works just fine where the distance between the two points is a few centimeters or kilometers, or even infinity.

But in the case of a quantum mechanics "single-point charge" where the distance between the charge and itself is evidently zero, there is a big problem. The formula fails for a single particle because you would have to divide the result by zero. As we know, you cannot divide something by zero. To "divide" something means to separate it out into parts. You can divide a pie into eight slices or five and a half slices or 1000 slices. But you cannot "divide" it into "zero" slices because that is not dividing.

Another way to explain this is that when you try to divide something by a very small number (very close to zero), the answer approaches infinity; and it is clearly provable that the energy of a single-point charge does not approach infinity just because it has come near itself.

This is of great interest to us because this relationship happens to be one of the foundations upon which almost all the great advances in electrical energy were constructed and is extremely relevant to those seeking Energy Miracles. (See the second key to the Energy Miracles.)

According to quantum mechanics:

1. A photon or electron is a point charge (an electric charge concentrated at a single point without mass or space).
2. The amount of energy located in the field around that photon or electron is represented by an amount that must be divided by zero.

The failure of the formula when the distance between the two points is zero proves that the preceding two ideas cannot both be true. And if a photon is not a point charge, then what is it?

Whatever it is, this section provides strong evidence that a single photon or electron has no resemblance at all to the picture of "quantum" energy that has been painted for us by the quantum mechanics for the past 100 years.

Richard Feynman suggested the failure of quantum mechanics to address and solve this difficulty throws the entire quantum theory into doubt.[64]

The answer to this seeming conundrum lies in the fact that two terminals are required to produce energy. This is observable in magnetism, electricity, and gravity and yet ignored by the quantum mechanics.

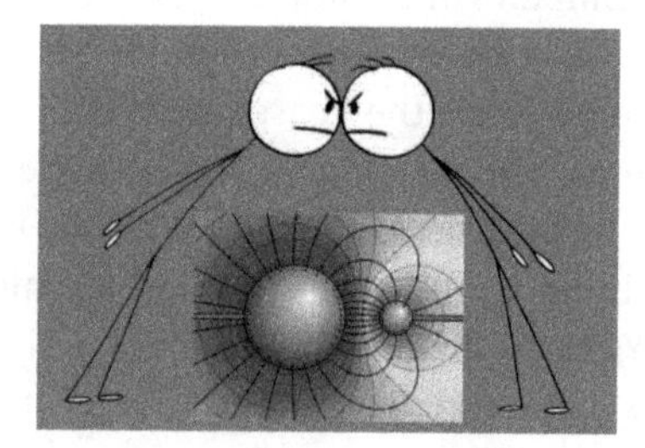

Figure 73 Two particles are needed for energy to be created.

Quantum End of the World: Entropy

The quantum mechanics have their own version of the apocalypse (the idea that the end of the world is near). It is called entropy.

---

[64]Richard Feynman, *The Feynman Lectures on Physics*, Vol. 2, Lecture 8, *Electrostatic Energy*.

Figure 74 Entropy: quantum mechanics end of the world.

The dread in which college students often hold their thermodynamics classes will be found to stem in greater or lesser degree from failed attempts to grapple with the concept of entropy. It is not their fault. Their professors mostly did not understand it either. Searching Google and physics textbooks for a good definition or means to teach it is pretty much a fruitless endeavor. Entropy was the first scientific concept to be appropriated by the quantum mechanics, and ironically, it came out of 19th-century efforts to make coal-burning steam generators more efficient.

In 1827 Robert Brown was looking through his microscope at particles trapped in water and noticed these particles moved around randomly. It was later noted that when the water was heated up, the activity of these particles increased and they even emitted radiation. How and why this "Brownian motion" occurred was unknown. Other scientists at the time, unable to explain it, made various assumptions about it, but its cause remained a mystery. Max Planck became involved, and in so doing became an inadvertent and unwilling founder of quantum mechanics. Planck, as discussed earlier in this chapter, was an honest scientist looking for broad principles that would explain things, not speculative meanderings at the atomic level that would not. In particular, he was looking for an electromagnetic foundation for thermodynamics, and in his opinion, a deeper explanation of thermodynamics could only be found in the dynamics of a medium such as an electromagnetic aether. (The concept of an aether medium has always been a vital element in the subject of energy creation, and we take it up fully a little later on.) For now, let us just say Planck had set a difficult task for himself and, as mentioned in the previous section, he turned for help to Ludwig Boltzmann's mathematics. The Boltzmann–Planck equation estimates the entropy of a system in terms of the total number of possible microscopic actions of the individual atoms and molecules in the system, making it unnecessary to count those water molecules or examine any one of them individually to be able to determine the overall conditions of the system such as its temperature and pressure.

For example, when water is heated up to steam, you suddenly have billions of water molecules floating around in a gaseous state. The number of increased particles in motion and its attendant pressure is its entropy. There are things you can know about the entropy of a system and things you might not be able to or might not need to know about it. You might need to know the quality of the water before heating, but you probably would not need to know whether the water came from a river or a lake. You would also never know *exactly* how many water particles were involved in a specific heating process, but then again, neither would you have much need of that information. It would be enough to know that if the water contained in a turbine of a specific size was heated to a certain temperature, you could compute the specific amount of pressure that could be developed.

Brownian motion was one of the great discoveries of the 19th century. Boltzmann's mathematics, predating quantum mechanics by many years, brought increased understanding and control of the steam systems used to generate electricity, but it is far from universal in application. Einstein entered confusion into this concept with his

1912 paper on the subject, when his predictions were significantly contradicted by many experiments and observations.[65],[66]

Entropy is an alluring enough concept to grab one's attention, and black enough to discourage optimism about the future of anything. Even Al Gore got caught as can be seen in the beginning of his book *The Future*[67] where he devoted an entire section to the subject:

> *"Entropy... causes all isolated physical systems to break down over time and is responsible for irreversibility in nature. For a simple example of entropy, consider a smoke ring: it begins as a coherent donut with clearly defined boundaries. But as the molecules separate from one another and dissipate energy into the air, the ring falls apart and disappears."*

This is a dismal commentary if you try to apply this principle to the world at large; and, thankfully, a false one. Gore then recounts his meeting with Nobel Prize recipient Ilya Prigogine. Prigogine had formulated a theory that was meant to alleviate the pessimistic inevitability of entropy. It is called the "order out of chaos" theory and posits that when an open system[68] is reaching the end of its cycle, with maximum entropy and maximum disorder, it not only breaks down, but the system then reorganizes itself into a higher form. In other words, an overall order spontaneously (miraculously) arises from the disorder with no need of control from any external agent.[69]

What is wrong with modern theories of entropy is their disregard for causality. Omitted by those who dreamed up the smoke ring in Gore's example is the fact that the ring was caused by something. It did not just happen. A guy bought a pack of cigarettes and a lighter, lit one, and inhaled a decent quantity of smoke. Then, with his mouth

[65]S. B. Brush, A history of random processes: I. Brownian movement from Brown to Perrin, *Archive for History of Exact Sciences*, **5**, 1, 1968.

[66]G. H. Pollack, *The Fourth Phase of Water*, Ebner & Sons Publishers, Seattle, WA, USA, 141–147.

[67]A. Gore, *The Future: Six Drivers of Global Change*, Random House, New York, 2013.

[68]An "open system" is one that interacts with other systems. Some examples are the human body, your family, your company, and the solar system. In the real world, all systems interact with other systems.

[69]To be fair, Prigogine also discovered that by injecting energy into a system, the inevitable maximation of entropy imposed by the second law of thermodynamics could be reversed. Stated another way, bad things can be reversed if you become cause over them. For that, he won the Nobel Prize in Chemistry.

and tongue set just so, he exhaled a short puff of that smoke into the surrounding atmosphere. The relative heat, humidity, gravity, and other cohesive elements within the smoke kept that ring intact against opposing forces from the environment. For a while, the strength of the initial creation was sufficient to keep it round, but as counter efforts from the environment started to overwhelm it, its shape and character began to dim.

If you tried blowing a smoke ring into a hurricane, you would not succeed for even a second. Creating a smoke ring in a perfectly still room would give it greater longevity. If you could blow a smoke ring into a vacuum with a complete absence of counter efforts, you might have it persist for a very long time. And if you could find a way to continue recharging the smoke ring with smoke, it might continue forever.

With the modern quantum concept of entropy, we have a smoke ring of doubtful creation, disappearing into nothingness in an "inevitable," yet un-caused, process, only to have grand new forms "spontaneously" emerging out of its destruction. Not a single causative agent can be found anywhere in the process, yet it is called a "law of nature." Only in states of insanity or criminality can you find so many no-bodies causing so many no-things.

In classical physics before 1900, the term entropy made some sense. The word is derived from "en," a Latin prefix used to specify something brought into a specific condition, and "tropos," a Greek word describing the motion of matter as it moves through different states of existence. In other words, entropy referred to the change resulting from a physical process such as the application of heat. Example: No heat: very little entropy in a quiet puddle of water. Lots of heat turning water to steam: lots of entropy as the steam expands out as gas to fill the available space.

The quantum mechanics replaced the original meaning with arcane definitions such as: "a measure of the unavailable energy in a closed thermodynamic system that is usually considered to be a measure of the system's disorder"; and "the degradation of matter and energy in the universe to an ultimate state of inert uniformity."[70] Who can blame the apathy of a physics student who has been required to make sense of such drivel?

---

[70]Merriam-Webster Dictionary

Quantum mechanics will tell you that entropy is the supreme law of nature and "proves" that disorder and chaos always increase over time, that **left to its own devices**, life will always become less structured, sandcastles will always get washed away, weeds will overrun gardens, ancient ruins crumble, and cars rust. The key is in the phrase "left to its own devices." There is no such thing in our world. There are causes, and there are effects of those causes.

Cause and effect and the action cycle

There are many **action cycles** in the universe: Something starts, then continues for a while or changes, and then ends.

A fellow gets hired at a new job (**start**), works 5 days a week for 3 years (**continues**), and gets fired one day for continual lateness and sloppy work (**stops**). Another fellow gets hired at the same company on the same day (**start**), works 5 days a week for the same 3 years (**continue**), and is then promoted out of his job into an executive position (**stops**). Both are action cycles. The cause of the first guy being fired was his sloppy work and refusal to be punctual. The cause of the second guy's promotion was his contributions to the success of the company plus his willingness to work hard. There is nothing inevitable about either one of those things. It is cause and effect. The action cycle "start–change–stop" shown in Fig. 75 operates in parallel with "cause–distance–effect."

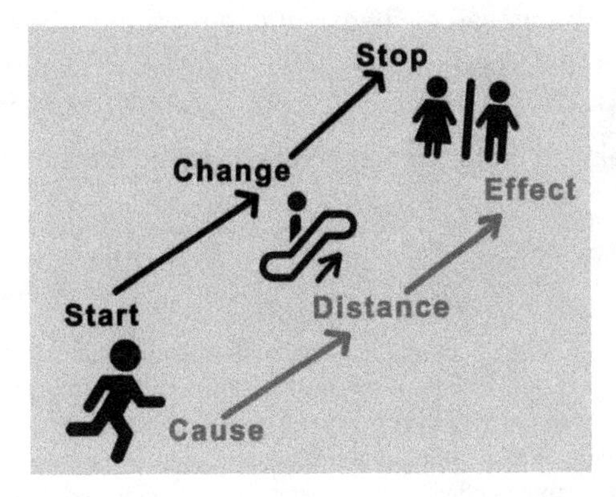

Figure 75 Relationship of start–change–stop and cause–distance–effect.

Two families both buy a new car of the exact same model. One of the cars gets continuously dented, dinged, and scratched up from inconsistent and careless driving, and left outside to face the elements. The other car is more carefully driven, kept free of salt deposits, and housed in a garage out of the glare of direct sun when not in use. It is no surprise that one will be rusted out after a few years, and the other will not. It is cause and effect. It has nothing to do with a hocus pocus entropy.[71]

What people generally refer to as magic are just events where the cause is not obvious. When an elephant appears magically on a stage, you can count on there being an actual cause that has been kept out of the sight of the audience. (The magician's day job was that of a carpenter, and he spent 5 days constructing cranes, well-oiled trap doors, and fake walls to pull off that illusion.) Just because someone cannot immediately spot the actual cause does not mean it is not there.

As shown earlier, the world seems hell-bent on perpetuating the use of 19th-century energy systems to service its power needs. This is largely because scientists in the early 20th century redefined physical processes like entropy to omit causes and effects. This has everything to do with our search for a 21st-century energy alternative because energy consists of causes and effects. Take lightning. It starts with a difference in electrical potential being created in the clouds. It continues as the lightning strike travels from the cloud to Earth. And it stops at the explosive end of cycle when it hits the Earth. Instead of theorizing about things they cannot see, scientists would do better to sit down and study the different types of energy—there are only three—that are discussed later in this book. They also need to become experts on the mechanism of cause and effect.

---

[71]The key to longevity in an action cycle lies in the **continue** part of the cycle. In the example of the guys with the jobs, one was continuously creating his job and the other had ceased creating it. In the car example, one family was continuously creating the car and keeping it free from counter efforts in the environment. The other car was made the unfortunate effect of the owner's poor driving and damaging weather conditions. No inevitable entropy was involved.

Energy, the Speed of Light, and Albert Einstein's House of Cards

> *"I believe that I have really found the relationship between gravitation and electricity, assuming that the Miller experiments are based on a fundamental error.* ***Otherwise, the whole relativity theory collapses like a house of cards.****"*
>
> —Albert Einstein, in a June 1921 letter to Robert Millikan fearing the results of Dayton Miller's experiments to prove the existence of aether.[72]

The 19th century saw the Golden Era of Electricity heralded by the discovery of the electron and the formulation of Maxwell's laws linking electricity with magnetism. James Clerk Maxwell was one of the towering figures in science, with his greatest achievements in the subject of electromagnetic theory. For all of recorded history, humanity has observed electrical effects in nature such as lightning and the attractions caused and sparks emitted when certain objects were rubbed together. As for magnetic effects, Christopher Columbus brought a magnetic compass with him on his voyages to discover the New World, but as far back as the year 2637 B.C., a Chinese emperor had already constructed a chariot carrying a prominent female figure that always pointed to the south no matter which direction the chariot was moving.[73] Maxwell's realization of the close relationship between electrical and magnetic phenomena sparked a great surge forward in electrical science and engineering. Maxwell was also the first to draw an analogy between liquid waves and lines of electric and magnetic force—and then prove them with mathematical formulas. His electromagnetic theory stands as one of the greatest unification triumphs in the history of science, and it was based firmly on the concept of an aether medium through which all electromagnetic waves (including light) traveled. This changed in the early 1900s with the relativity theories of Albert Einstein.

---

[72]Ronald W. Clark, *Einstein: The Life and Times*, World Publishing Co., 1971.

[73]Paul F. Mottelay, *Bibliographical History of Electricity and Magnetism*, Charles Griffin And Company Limited, 1922.

Speed of light in a vacuum and Einstein's relativity theories

Einstein's **Special Theory of Relativity** was released in a paper he wrote in 1905 and is represented by the best known formula in all of science: **$E=MC^2$**.

It is relevant to our subject here because the *E* stands for energy and we are in search of an Energy Miracle. This theory, Einstein's attempt to explain how space and time were related, was based on a "thought experiment"[74] unrelated to laboratory result or observation. It was premised on the assumption that the speed of light did not vary. Einstein originally stated that the speed of light was constant at all times and places, no matter from where it comes, what it travels through, or who is looking at it. Of course, that is not true. For one thing, visible light is measurably slowed down in transparent media such as air, water, and glass.

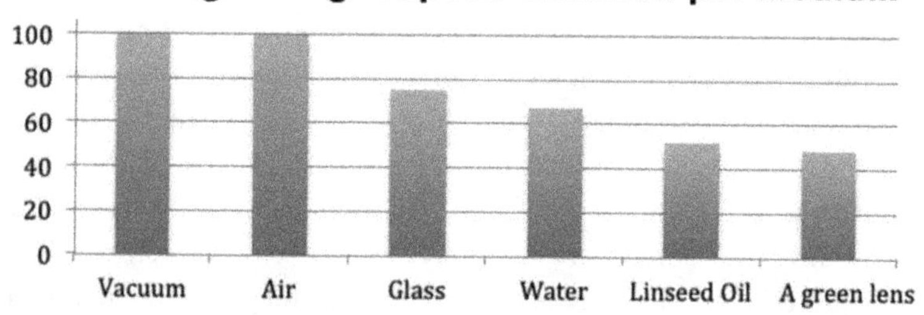

Figure 76 Light changes speed in different mediums.

Other electromagnetic waves (such as X-rays and gamma rays) travel better through solid matter but are also slowed down when doing so. To make his relativity theories work mathematically, Einstein was forced to make a revision. To overcome this "speed of light" obstacle, he advanced the concept of "speed of light in a vacuum" (a vacuum being defined as a space with absolutely nothing in it to influence or interfere with the light). A vacuum would not only guarantee Einstein his constant speed of light, but it would relieve him of having to worry about the initial impulse of the light

[74]See the section "Scientific Experiments and Thought Experiments" in this chapter.

at its origin or interference along its route. Light could just be Light. Period!![75]

This light-in-a-vacuum idea was a convenient solution that was widely embraced, even though no such vacuum exists. Vacuum speaks to the relative density of particles in a space: less particles = more vacuum. But even modern quantum mechanics now admit that outer space is not really empty. They acknowledge that the "quantum vacuum" is a soup full of tiny particles that are blinking on and off at a great rate and effecting the speed of light.[76] (More on that shortly.)

The "speed of light" has assumed an almost godlike status in energy science, one that can never be even vaguely criticized. Daring to do so has ruined the reputation of great scientists such as Dr. Dayton Miller, as we shall soon see. In fact, many things are known to influence and change the characteristics and speed of light, and because this is central to the search for Energy Miracles, we discuss a few of them here.

Initial impulse and interference create variations in lightspeed

Figure 77 shows examples of how the initial impulse of energy and the medium through which it travels will affect both its characteristics and speed.

Gravity creates variations in lightspeed

Besides the two conditions in Fig. 77, it has been well measured that large masses such as black holes and large stars deflect light. So does the magnetic force of the Earth.

---

[75]Some will argue that glass and water are not vacuum. True, but the irony is that quantum mechanics is willing to allow the speed of light in air to be considered relatively the same as that in a *vacuum*. This despite the fact that air on Earth is chock full of gases, particles, and ions, as well as being much warmer, more humid, and heavily influenced by both the planet's gravitational and magnetic forces. The air in the Earth's atmosphere is far from a vacuum and has a noticeable influence on light.

[76]M. Urban, The quantum vacuum as the origin of the speed of light, *European Physical Journal D*, **67**, 2013, p.58.

Figure 77 According to Einstein, the speed of light is constant. But what of the power and type of impulse that originated it? And what about the interference it encounters along its path?

Figure 78 shows some ways the speed and character of light are affected in the "vacuum of outer space" on its way to Earth. "A" shows relatively unhindered beams of light. "C" shows the alteration of light's direction, speed, and wavelength when its path takes it close to the gravity field of a large star. "B" shows light being totally stopped when a black hole lies directly between the light source and the Earth.

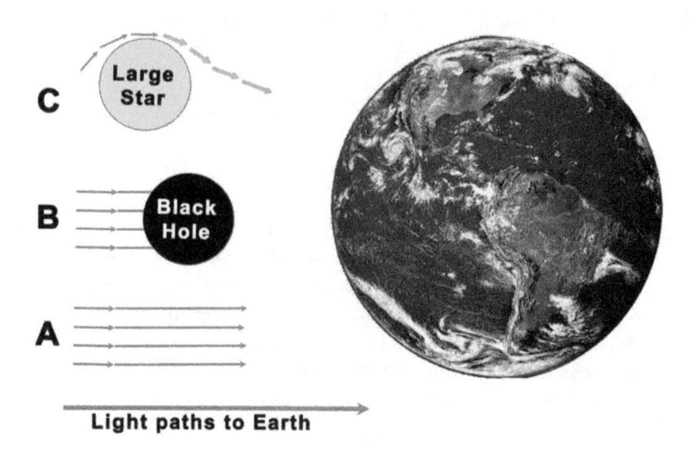

Figure 78 Light speed being altered in a "vacuum."

The absolute cold in outer space creates variations in lightspeed

Incontrovertible evidence to disprove the assertion that the speed of light traveling through free space never changes comes from a Danish physicist leading a combined team from Harvard University and the Rowland Institute of Science that succeeded in slowing a beam of light from 300 million m/s down to just 17 m/s simply by lowering its temperature.[77] The irony here is that the authors did not notice that their findings contradicted the immutability of Einstein's speed of light. Why? Because "their test wasn't done in a vacuum—it was done in a laboratory." Yes, but does not it follow that the severe temperatures of interstellar space would have a similar effect on light's speed? When the speed of light is measured, it is measured traveling through space. The average estimated temperature in space is −270.15°C, just a few degrees shy of the absolute zero point where all activity and motion cease. That is the exact temperature that was used in the aforementioned Danish and Harvard lab experiment, which was proven to so drastically reduce the speed of light.

Earth's atmospheric layers influence lightspeed

There are measurable changes in speed and direction when light gets into the Earth's atmosphere. The global GPS network spends much of its time compensating for those changes. The ionosphere is the largest source of error for standard GPS and NOAA posits the effect is a function of the density of the electrons in the atmosphere.[78]

Mapping the heavens

If the speed of light can be measurably altered within the few hundred meters of the Earth's atmosphere, how much more vulnerable to change is a wave of light traveling a couple of hundred billion kilometers through space? Any pilot knows that the current

[77]L. V. Hau, S. E. Harris, Z. Dutton, and C. G. Behroozi, Light speed reduction to 17 metres per second in an ultracold atomic gas, *Letters to Nature*, **397**, 1999, 594–598.
[78]NOAA posits these changes to be the effect of the density of the atmosphere. See M. Codrescu, The influence of the ionosphere on GPS operations, *Applications of GPS/GNSS in NOAA*, Boulder, Colorado, October 24–25, 2007.

wind conditions, updrafts, lightning, clouds, and the presence of other aircraft all influence the path and travel time of a plane from Point A to Point B. Knowing only the average history of a particular flight path is not enough. The same holds true with GPS applications. In order to come up with an expected time of arrival, the technology of Earth's GPS network requires a continuous input of variable and varying data, including current location, map coordinates, destination location, estimated distance to the destination, history of the conditions that have previously been known to exist on that route, new conditions that may emerge, average driving speed of the specific driver, current road conditions, and others. The time spent on the interstate highway is part of it. The time spent waiting for lights to change on crowded downtown streets is another part of it. These are all averaged to achieve a "time of arrival."

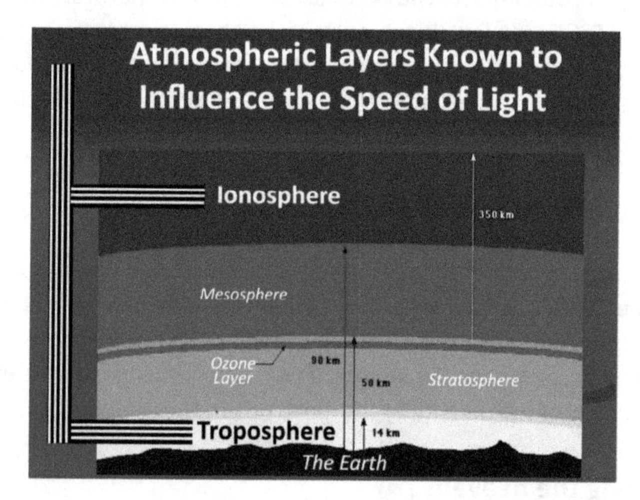

Figure 79 The troposphere (out to 17 km) and the ionosphere (out to 350 km) are both known to interfere with electromagnetic energy and require continuous compensation in Earth's GPS network.

So it is with light. In order to accurately figure how fast light is traveling, you would have to map the heavens on its route and determine whether or not there is an interfering black hole, or dense masses of gases or particles, or variations of extreme temperatures, or gravitational pull from other stars.

Lightning does not travel at the speed of light

Lightning is another example of well-defined variations in the speed of light. Lightning discharges contain the entire spectrum of electromagnetic radiation (from shockwaves, to visible light, to X-rays and gamma rays) and has been many times measured to travel one-third to one-half the speed of light. Even more telling is that the luminous continuously propagating sections of a lightning strike, called Dart Leaders, have been measured with high-speed cameras to travel only **one-thirtieth** the speed of light.[79] Which simply means that light speed varies as a function of both the type of light and the interference of the medium or mediums through which it travels.

Accuracy of currently averaged speed of light

What is the "speed of light" anyway? Before Einstein and the quantum mechanics, the speed of light had been measured by Leon Foucault in 1862 by projecting light beams onto rotating mirrors. He computed the speed of visible light as *approximately* 298 million m/s. Foucault's classical experiments were accurate enough to prove that light moved more slowly in water than it did in air. Using the same principle but with different (ostensibly more accurate) equipment renders slightly different results. In 1983, 120 years later, an international committee on weights and measures set the speed of light in a vacuum at 299.7 million m/s. The same committee then defined the meter as 1/299.7 millionth of the distance light travels in one second. It is circular, like answering the question, "How fast can a car travel?" with, "It can make it to Omaha in one day." And then determining the distance to Omaha with the logic, "It's the distance a car can travel in one day." It is certainly convenient, but does not account for accidents on the highway or souped-up eight-cylinder engines. And, notice the difference between our 21st-century speed of light and Foucault's 19th-century speed of light is only about four-tenths of 1%.

The quantum *constant speed of light in a vacuum* is clearly cracking. But our interest in all this is only its effect on the search

---

[79] J. R. Dwyer and M. A. Uman, The physics of lightning, *Physics Reports*, **534**, 2014, p.153.

for alternative energy sources. Up until Einstein's revelations in 1904, just about all of science was utilizing a principle called the aether medium to explain how light and energy traveled from one place to another. This medium did not just interfere with light; it actually *enabled* it. The function of some kind of aether medium had been factored into literally every landmark discovery in electrical engineering. Einstein himself had endorsed the aether up until he realized it challenged his theory that the speed of light was a constant. His solution was elegant: *reject the aether*. He reasoned that if his theory were correct, there was no room for an aether in science. Because this single action has had such a profound effect on energy science, we examine it closely in the next few pages.[80]

The Aether

> *"All attempts to explain the workings of the universe without recognizing the existence of the aether and the indispensable function it plays, are futile and destined to oblivion."*
>
> —Nikola Tesla

Since antiquity, the doctrine has existed that force cannot be communicated except by pressure or impact, and one body can only influence another if they are directly or indirectly connected. Punch an enemy or kiss your girlfriend—in both cases force is transmitted through direct contact. Throw a ball through a window;

---

[80]Aside from the speed of light problem, Einstein realized his initial relativity theory suffered from another shortcoming: It failed to encompass the effects of gravity. So, after 10 years of further contemplation, he released his General Theory of Relativity in 1915. As with his earlier Special Theory of Relativity, this one was born out of thought experiments. In one of the key ones, he imagined being in an enclosed elevator accelerating up through the vacuum of space. In another, he imagined space to be a giant trampoline. If you placed a large bowling ball on it, the trampoline would stretch and if you then rolled some small billiard balls out on it, they would all roll toward the bigger ball. In this theory, Einstein tells us that what we perceive as the force of gravity has nothing to do with the relative mass of the objects involved (such as the huge Earth and the tiny apple), but arises solely from the "curvature of space and time." Though an easily visualized analogy, nowhere has quantum mechanics ever shown how the geometric abstraction called "curved space-time" is able to create a gravitational wave that propagates at the speed of light.

the damage is caused by direct contact. An explosion on one side of a lake causes a huge swell on the other side of the lake. The water connecting the two sides of the lake transmits the force. Similarly, sound is transmitted through the medium of the air. These examples demonstrate that space is neither a void nor a vacuum, and there must always be something between the cause point and effect point that connects the two and does the impacting. This principle applies to bowling balls and bazookas, to incense and industrial pollution, and is behind the transmission of electromagnetic waves of all sorts, including starlight, sunlight, X-rays, and cosmic rays. It also lies behind the forces of magnetism and gravity. Things do not happen at a distance without some intervening contiguous medium between the cause point and the effect point.

Aether (also spelled ether) coming from the Greek aithēr, from *aithein* to ignite, blaze)) is the name long ascribed to that medium. In the Middle Ages, aether was thought to describe a special substance that filled the heavens. In later times, when a word was needed to describe a substance that filled the void between planets, it was the logical go-to word. Aether as a term is, like its substance, exceedingly flimsy. No one up to now has been able to figure out what it is composed of or measure it in any meaningful way. It may be just a piece of gossamer, but such a gossamer powerful enough to tie half the brain of science behind its back.

More importantly to our story, the concept of an aether medium has been at the foundation of every major **electrical** advance in history as per Table 2 given earlier in this book.

Aether before Einstein

For over 2000 years, up through the beginning of the 20th century, the aether concept was fundamental to science. It was used by Aristotle in ancient Greece and Lucretius in ancient Rome. Jewish, Christian, and Muslim scholars in the Middle Ages discussed it. The existence of aether was confidently used by Sir Isaac Newton to explain both gravitation and light, and he insisted that it was impossible for such interactions to work without it. Leibniz concurred. The existence of an aether was reinforced in the 19th century after Augustin Fresnel and Thomas Young made it obvious that light must be treated as a wave and the aether was recognized as the transmission medium for all electromagnetic waves, including light. The works of Charles Augustin de Coulomb, Benjamin Franklin, Luigi Galvani, Allesandro Volta, André-Marie Ampère, George Ohm, Michael Faraday, William Thomson (Lord Kelvin), James Clerk Maxwell, Heinrich Hertz, Hendrik Lorentz, Thomas Edison, and Nikola Tesla were all based upon the presumption of a transmission medium they called "aether." Even Einstein, up until 1904 believed in an elastic aether in which light was propagated with varying velocities.[81]

Just because all the above used the same term, we should not think they all agreed about what the aether consisted of. In the history of physics, rarely has a term been imputed with such a huge number of different meanings, or so many guesses put forward to explain its structure. At one point in the 19th century, there were 14 different concepts of aether simultaneously being used.[82] Despite those differences, all agreed that the aether was a medium, a "something" that either filled or existed in space, which was responsible for bearing or transmitting electromagnetic, gravitational, and other interactions. And there need not be a universal medium. The effect of the medium on a beam of light passing near a star (with its millions of degrees of heat) would be different from the effect on that same beam of light when it was traveling through interstellar space at temperatures of –270°C.

The mistake science made with regard to aether was to get hung up on its structure. Scientists could have just acknowledged that they did not **yet** have a clue as to its composition and left it at that. Of far more importance was the **function** of the aether. So far as is known, all waves (including electromagnetic waves) require a medium to propagate. This is something that seems to apply to all energies and particles: For electricity to travel down a wire, you need the wire. For lightning to travel from the clouds to the Earth, there needs to be some plasma-like conducting medium. For sound to travel from an explosion to your ear, some medium needs to exist to transmit the sound waves. The same applies to radio waves and the waves of light emitted by the sun and the stars.

As we have seen, the notion that this would have to apply to light gave Einstein a big headache. An aether existing in space would

[81]Ludwik Kostro, *Einstein and the Ether*, Apeiron, Montreal, 2000. p.6.
[82]IBID p.iii

have called into question Einstein's theory that all light traveled at a constant speed. A variable speed of light would have sunk his theory of relativity.

It is really a pity to pin all the bad stuff on Einstein for he was not a bad guy. Unlike many of the other major actors in the formative days of quantum mechanics, Einstein stood up to Hitler and then when Hitler was defeated, he also stood up to the other militarists who wanted to add nuclear weapons to their arsenals. Most of the worst effects he caused were inadvertent, and he regretted many of them up to the end of his life. Still and all, he must take his fair share of responsibility for the ensuing debacle in science. As far as energy discovery is concerned, arguably the most criminal action of the quantum mechanics was the abolition of the concept of an aether medium. Einstein, almost single handedly, in the effort to protect his theories, removed from science the concept of energy propagating thru a medium.[83]

Measuring the aether

Despite all the interest in it, no one has ever been able to measure or perceive actual aether particles. I refer you to the 1000 pages of Whittaker's exhaustive *History of the Theories of Aether and Electricity*[84] and the more recent compilation *Conceptions of the Ether*.[85] There have been about as many theories about the composition of the aether as there have been those who have given it serious thought. Did the particles have mass or not? Were all the particles identical to each other? Were they round or some other shape? Did they spin? Did they have different colors? Were

[83]Einstein was an adherent of the concept of aether up to 1905. Then, for 11 years he rabidly denied its existence. In 1916, in some confusion, he reintroduced a kind of non-aether aether definition to his theories, and then on May 5, 1920, in an address delivered in the University of Leyden entitled *Ether and the Theory of Relativity*, he said "According to the general theory of relativity, space without ether is unthinkable; for in such space there would be no propagation of light." But by then, it was too late. The quantum mechanics had grown new legs, and these legs ran in a vacuum unencumbered by any medium or aether.

[84]Sir Edmund Whittaker, *A History of the Theories of Aether and Electricity*, Philosophical Library, New York, 1951. Vol. I The Classic Theories; Vol. 2 The Modern Theories 1900–1912.

[85]G. N. Cantor and M. J. S. Hodge, *Conceptions of the Either: Studies on History of Theories of the Ether 1740–1900*, Cambridge University Press, 1981.

the particles in motion or not? Did they have attractive or repulsive forces? Did they come in pairs or triplets? Was there space between the particles or was the aether some kind of continuous medium? Was the medium solid or liquid? Did the aether have elastic qualities or did it transmit impulses in some other way? If composed of particles, were they randomly scattered with different densities and properties in different locations or was there some cosmic blueprint or template?

At the beginning of the 20th century, even while this conjecturing was ongoing, every major scientist assumed that the existence of an electromagnetic "wave" required some medium through which it was propagating. Water waves are caused by an impulse or particle, but without the water, there would be no wave. Similarly, without a medium through which to travel and create an effect, there would be no heat, no light, no lightning, no sound, and no gravity. When James Clerk Maxwell showed that electricity, magnetism, and light were just different manifestations of the exact same phenomena, he was unequivocal that electromagnetism was dependent on the fact that, "*there is an aethereal medium pervading all bodies.... capable of being set in motion by electric currents and magnets, and this motion is communicated from one part of the medium to another by forces arising from the connection of those parts.*" [86]

By 1905, besides his speed of light problem, Einstein realized there were other ways the existence of an aether could invalidate his relativity theory. For one thing, if light was indeed a wave flowing through aether, then if you were moving through the aether toward the light source you should see the waves going by you at a faster speed than if you were traveling in the same direction as the waves. This is obviously true and had already been proven experimentally by the Danish astronomer Ole Roemer in 1676. While studying Jupiter's moons, he noted that their eclipses took place sooner than predicted when Earth was moving toward Jupiter and occurred later than predicted when Earth was moving away from Jupiter, meaning something was interfering with the speed of light. But Einstein and his supporters did not agree to allow this 225-year-old data to interfere with the extremely neat theory of relativity.

---

[86]James Clerk Maxwell, The dynamical theory of the electromagnetic field, *Philosophical Transactions*, 1864.

James Clerk Maxwell had suggested a method to detect the aether: "*If it were possible to determine the velocity of light by observing the time it takes (for light) to travel between one station and another on the Earth's surface, we might by comparing the observed velocities in opposite directions, determine the velocity of the ether with respect to these terrestrial stations.*"[87]

For sure, attempts were going to be made.

The Michelson interferometers

> "*I believe that I have really found the relationship between gravitation and electricity, assuming that the Miller experiments are based on a fundamental error.* ***Otherwise, the whole relativity theory collapses like a house of cards.***"
>
> —Albert Einstein, in a June 1921 letter to Robert Millikan fearing the results of Dayton Miller's interferometer experiments to prove the existence of an aether.[88]

It is worth repeating that quote because it refers to the most famous attempts at measuring the aether. These attempts occurred in 1887 and 1904 by Albert Michelson, Edward Morley, and Dayton Miller using a device Michelson had constructed, called an interferometer.

An interferometer is an investigative tool that takes advantage of the wave-like characteristics of light. By carefully observing the pattern made when two sources of light are merged, it can be ascertained whether or not something has interfered with that light (hence the name *interfer-ometer*).

The device, based on the design in Fig. 80, would receive and split a light beam down the middle at the Splitter, sending each half along separate paths to the mirrors shown as Mirror 1 and Mirror 2. The mirrors would reflect the two beams of light back to their source where the elapsed time would be measured. [89] Well, not

[87] T. Bethell, *Questioning Einstein: Is Relativity Necessary?* Vales Lake Publishing, Colorado, 2009.

[88] Ronald W. Clark, *Einstein: The Life and Times*, World Publishing Co., 1971.

[89] The experiment did not directly measure the time difference between the two light beams (such as with a clock). Instead, the interference patterns they made on a receiving board would be compared. Any difference would signify that the aether was causing a slowdown in light speed.

exactly "measured." Changes of light speed could only be *inferred* by differences in the resulting interference patterns displayed on the device's detector screen. They should be identical. Any change in the pattern between the two sources would indicate something had interfered. In the case of the Michelson, Miller experiments under discussion, that "something" was called the aether medium.

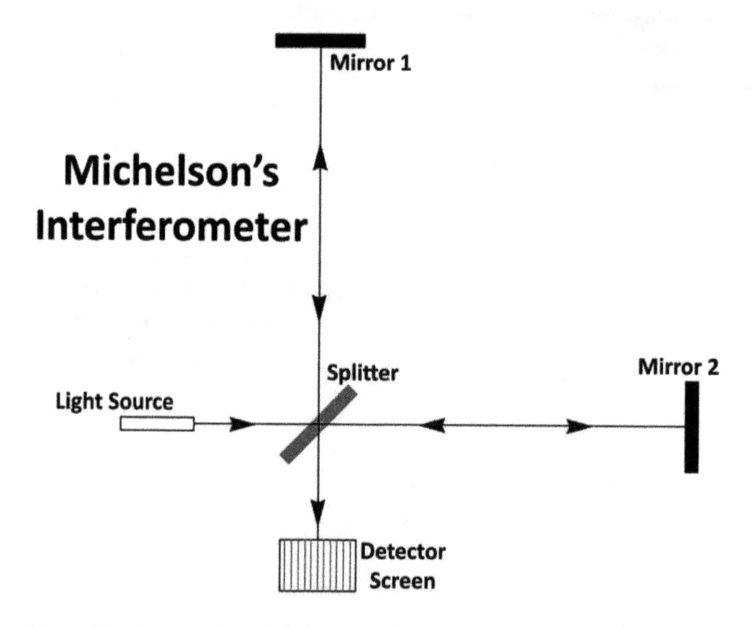

Figure 80 Design of a Michelson interferometer.

Michelson theorized that if he sent one beam of light in the direction of the Earth's rotation, and the other beam at a right angle to the first one, if there were an aether, it would interfere differently with the two beams causing a variation of light speed.

Google these experiments, or read about them in any physics text, and you will learn that they gave null results and were the most famous failed experiments in history. Only they did not and they were not. Michelson alone conducted the first experiments, but he later enlisted the aid of Edward W. Morley, and then Dayton Miller. When Michelson and Morley were too old to continue, Miller carried on the experiments and a study of Miller's papers shows the interferometer results demonstrated a definite difference in the two beams of light of between 8.0 and 8.8 km/s depending on the time of day the measurements were taken.[90],[91] This was not some casual or amateur observation. This was based on over 200,000 individual measurements taken over 20 years. Dayton Miller was no lightweight. A graduate of physics from Princeton University, he was appointed Chairman of the Physics Department of Case Western Reserve University, and later elected President of both the American Physical Society and the Acoustical Society of America. He was also a member of the National Academy of Sciences and served as Chairman of the Division of Physical Sciences of the National Research Council. And he wrote in 1928, "*The effect [of ether-drift] has persisted throughout. After considering all the possible sources of error, there always remained a positive effect.*"

Einstein did his best to discredit these experiments. After 1904, Dayton Miller continued to refine his test techniques and conducted hundreds of thousands of experiments on the aether. By the 1920s, he had developed the most sensitive and accurate interferometer in the world. And when he ran the test again in 1925, he recorded a difference in light speeds of up to 10 km/s. It was a positive sign of the change in light speeds due to the aether, and thus central to Einstein's relativity theories, but Einstein did not want to know anything about it. He made known his attitude toward data that might upset his theories of relativity in a 1925 letter to Michele Besso: "I have not for a moment taken them [Miller's experiments] seriously."[92] Why he would so cavalierly reject results of an experiment bearing so heavily on the central plank of his theories he made clear in another letter (this one to Edwin E. Slosson on July 8, 1925): "*My opinion about Miller's experiments is the following. ... Should the positive result be confirmed, then the special theory of relativity and with it the general theory of relativity, in its current form, would be invalid.*"

Dayton Miller was ridiculed, and his work was dismissed for suggesting a variation in the speed of light. But he subsequently obtained significant positive results in four more series of experiments. More recent aether-drift experiments from the last

[90]D. C. Miller, The ether-drift experiment and the determination of the absolute motion of the Earth, *Reviews of Modern Physics*, **5**, 1933 (available for download at www.energymiracles.net).

[91]D. C. Miller, The ether-drift experiment and the determination of the absolute motion of the Earth, *Nature*, 1934 (available for download at www.energymiracles.net).

[92]P. Speziali, *Albert Einstein–Michele Besso, Correspondence 1903–1955*, Paris, Guida.

quarter of the 20th and early 21st centuries, by Galaev, Munera, and others, using radiofrequencies, light-beam interferometry, and other novel methods, seem to provide additional substantiation of the existence of an aether medium in space.[93]

The interferometer results were in no case null, but they were ignored when the magnitude of the differences turned out to be less than originally predicted. Miller explained the diminished magnitude as owing to differences in the nature of the aether near the Earth compared to that in free space. Despite their positive results, those experiments were considered the first strong evidence against the then prevalent aether theory, and led to its abandonment. At the same time, they were said to give credence to Einstein's relativity theories. To those experiments and the interpretation of their results can be ascribed the end of the Scientific Revolution that began in the 11th century. It opened the door to quantum mechanics.

Modern GPS technology detects the aether

The purpose of those earlier interferometer tests was to detect the aether by noting differences in light speed from two beams of light. In 1900, light speeds could not actually be measured directly by a clock, because the clocks of the day were insufficiently accurate. One hundred years later, modern technology in the form of the synchronized atomic clocks of the GPS can be used to directly measure the speed of light beams circumnavigating the Earth.

The purpose of a GPS is to provide accurate locations (and times) to its user stations (and customers with smartphones) on the ground. The most important work goes on between the 24 GPS satellites that comprise the network. Each has an atomic clock on board. All the clocks are synchronized with each other, but this is not a one-size-fits-all synchronization. Radio signals between satellites in a vacuum or in free space should be relatively interference free according to Einstein. But this is far from the case in GPS networks. GPS software (and software engineers) spend much of their time adjusting the travel time of signals between satellites to compensate for factors that should not really exist per Einstein's theories of relativity.

[93]J. DeMeo, Does a cosmic ether exist? Evidence from Dayton Miller and others, *Journal of Scientific Exploration*, **28**, 4, 2014, 647–682.

For one thing, the GPS signals between satellites (and thus light) travel faster in a westerly direction than in an easterly. This is called the Sagnac Effect and is a function related to the rotation of the Earth on its axis. GPS networks rely on software to continually monitor the changes of speed and time and compensate for the errors this creates. Per Einstein's relativity, the Earth's rotation is not supposed to affect light speed,[94] and the most obvious cause of the anomaly is some type of medium (aether) that is interfering with that light. These measured changes in light speed corroborate the existence of Maxwell's aether medium.

Another factor is gravity. The tiny atomic clocks in the GPS satellites tick slower the closer they approach the humongous Earth. The effect of a large mass on a small mass (Newton's law) is unambiguous and very noticeable to any GPS system and is continually compensated for. Finally, as shown earlier in Fig. 79, both the ionosphere and the troposphere create measurable changes in light speed, which must be continuously compensated for by the GPS software.

Why is any of this important? Because when you look again at Einstein's quote at the beginning of this section,[95] any proven variation of light speed disproves his relativity theory and puts the concept of an "aether" back on the table. From the preceding paragraphs you can see there are many visible, repeatable examples of light speed variances. The fact that theoreticians will justify them with new bizarre theories does not alter the fact that light can and does change speed. So those on the quest for an Energy Miracle should not be bashful about incorporating the concept of an aether medium into their endeavors.

---

[94]Einstein's exact quote from his original paper on the subject: "The speed of light in vacuum is the same for all observers, **regardless of the motion of the light source or observer**." When too many exceptions to this rule were being found, quantum mechanics invented other theories to reconcile those differences such as "relativity of simultaneity" and "time dilation."

[95]"*I believe that I have really found the relationship between gravitation and electricity, assuming that the Miller experiments are based on a fundamental error.* ***Otherwise, the whole relativity theory collapses like a house of cards.***"—Albert Einstein, in a June 1921 letter to Robert Millikan fearing the results of Dayton Miller's experiments.

Double-Slit Experiment. Is Light a Particle or a Wave?

> *"I will take just this one experiment, which has been designed to contain all of the mystery of quantum mechanics, to put you up against the paradoxes and mysteries and peculiarities of nature one hundred per cent. Any other situation in quantum mechanics, it turns out, can always be explained by saying, 'You remember the case of the experiment with the two holes? It's the same thing'."*
>
> —Richard Feynman, Nobel Laureate

If the interferometer experiments drove the quantum car to the edge of the precipice, then it was the double-slit experiment that pushed it out into the void.

It was the inability of the quantum mechanics to account for the fact that light was clearly composed of particles, yet exhibiting the properties of waves, that Prof. Feynman was referring to in the preceding quote. Up until 1900, Newton's laws were regarded as the foundation of all of physics. His theories involved the concepts of space and time, force and mass, along with the fundamental assumption of a direct connection between cause and effect. Newton held that light was a flow of particles emitted from a luminous source such as the sun; but one that also embodied a wave process. In 1801, the English physicist Thomas Young unveiled an experiment to the members of the Royal Society in London that corroborated Newton's view. The simple operation, known as the double-slit experiment, demonstrated the wave-like nature of light. In the early 1900s, this experiment was causing great consternation among certain scientists who could not fathom how to embrace the idea that light could be both particle and a wave simultaneously.

Here is a drawing of the experiment such as appears in physics textbooks (Fig. 81):

When this experiment is done in its modern form, a laser (a) beams individual particles of light at two slits or openings (b and c) cut into an otherwise light-proof barrier (S-2). On the other side of the barrier is a Detector Screen (S-3) that records the arrival of the particles. In the diagram, the wavy "interference" pattern on the Detector Screen shows that the shot particles are creating waves.

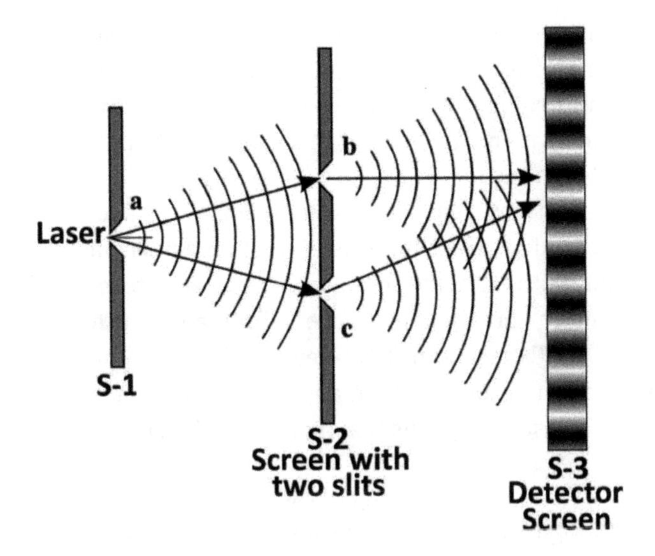

Figure 81 Young's double-slit experiment.

Einstein and other quantum mechanics, in the early 1900s, had decided that light was made up of tiny, indivisible units, or quanta, of energy, which they called photons, and Young's double-slit experiment created a problem for them: How could light be both a particle and a wave?

It did not take them long to come up with the following highly imaginative theory to explain it: When there was only one photon going through the apparatus at any one time, **the single photon must have been going through both slits at the same time and be interfering with itself.** And then, doing what they do best, the quantum mechanics came up with the mathematics to prove their imaginative theory. So, they could now say that light was neither a particle nor a wave but "something" that acted like both.

The thing is all these questions over which the quantum mechanics pondered so diligently admit of a much simpler, more logical, empirical and applicable solution.

Double-slit explained

When you send light particles through those two slits, you will get a wave pattern on the detector board. Though this confounded the quantum mechanics, a wave pattern is far from unusual or unique. When you put any kind of normal wave through those same two slits (such as ripples of water), you get the same wave pattern on the detector screen. (You can do this experiment in a bathtub with actual water waves.)

So what makes light so different? Einstein made it different. And his refusal to admit the existence of an aether medium made it different. But is it really so different? The scientists at the Laser Interferometer Gravitational-Wave Observatory (LIGO) project (operated by Caltech and MIT) have more experience observing the manifestations of light than anyone on Earth, and they pull no punches: "It just so happens that light waves behave just like water waves."[96]

Here are a few comments on the experiment that should make it more understandable:

All waves require a medium in which to wave, so if you assume the existence of some kind of medium permeating the space between S-1 and S-2 and between S-2 and S-3 in Fig. 81, wave patterns on the detector board would be not only predictable, but inevitable.

A double-slit experiment, conducted by Robert Austin and Lyman Page at Princeton University in 2010, confirmed this. They used a light source and a camera capable of detecting individual photons. The slits in Young's experiment were replaced by a thick set of filters, capable of limiting the flow of photons onto the screen. What was different was the way they timed their photographs.

After 1/30th of a second, you could see three distinct particles had landed. After a full second, you could see more particles arriving and there is already a pattern beginning to emerge. And after 100 seconds, the wave pattern has fully manifested itself. It can be empirically observed in this Princeton experiment that light is composed of individual particles, and when numbers of these particles are moving together, they act as a wave. A more important deduction, though, would be to compare the pattern shown in the bottom screen of Fig. 82 with that in Fig. 83.

---

[96]FIGO, *What is an Interferometer*, https://www.ligo.caltech.edu/page/what-is-interferometer.html

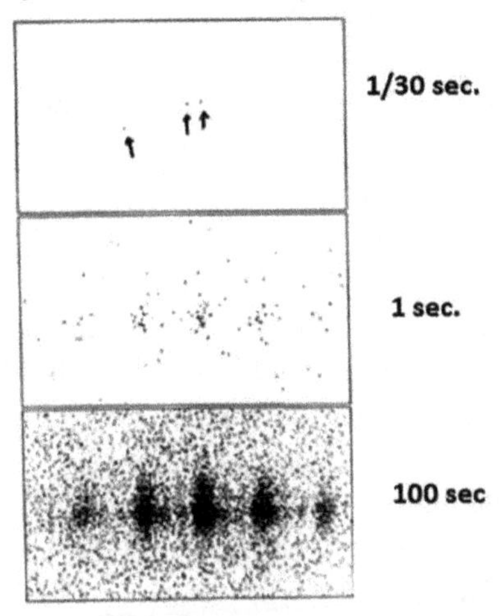

Figure 82 Lyman Page Princeton experiment.

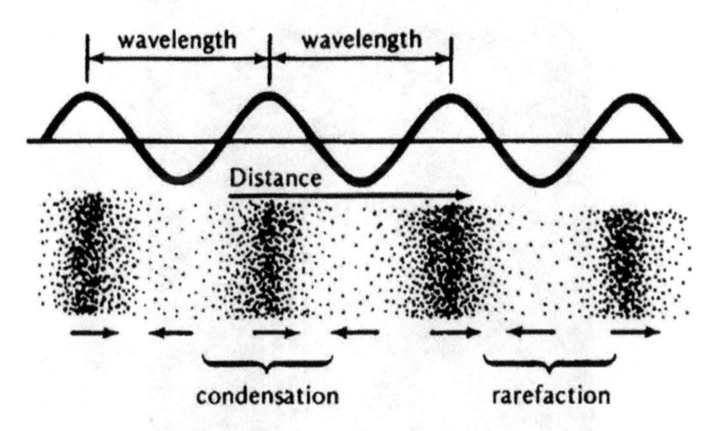

Figure 83 Condensation/rarefaction wave.

What you see here is a classic condensation–rarefaction wave, still used in other realms of science but denied to energy researchers ever since the quantum mechanics precipitously labeled it taboo.

To synthesize or not to synthesize? The wave/particle question

To paraphrase Richard Feynman, when considering scientists, there are them who synthesize and them who do not. Almost every great scientific achievement has occurred when a scientist discovered that two phenomena previously thought to be totally dissimilar were in fact different aspects of the same thing. Such advances (they are called syntheses) are seen in the works of the truly great scientists such as Galileo, Newton, and James Clerk Maxwell. Quantum mechanics are mostly them who do not. They are content to stick with their particles.

The entire debate surrounding the question, "Is light a wave or a particle," is a red herring, wrong target. See this photo in Fig. 84 taken by Ansel Adams.

Figure 84 Photo by Ansel Adams.

Is it a black photograph with white highlights? Or is it a white photograph with black shadowing?

Just as both black and white are intrinsic qualities of the photo, so does light embody both particles **and** waves. Take either the white or the black out of the preceding photo and you have nothing. Take either the waves or the particles out of the phenomenon of light and you similarly have nothing.

In Fig. 85 we can see an example of wave/particle duality. The particle on the left of the photo is a boat moving from right to left and creating a wave on a rather calm surface of water. When a particle causes a disturbance when moving through a medium, it is senseless to ask, "Is it a wave or a particle?"

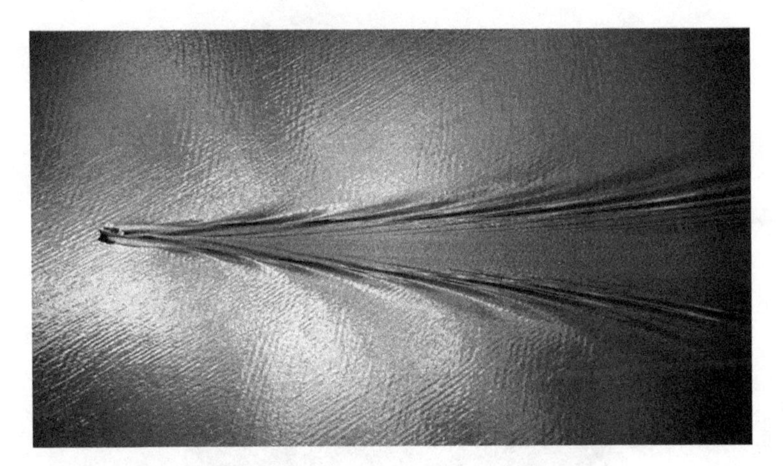

Figure 85 A boat making waves.

Two of the most famous of the quantum mechanics (Einstein and Schrödinger) spent the better part of their lives making the point that in order for quantum mechanics theory to be considered "complete," it would have to grow to embrace both of these aspects of light, not negate them. Until it does, quantum mechanics will continue to have no greater relevance than a coin with only one side.

Heisenberg's Uncertainty Principle and Its Opposite

"We can never know anything."

—Werner Heisenberg, *Physics and Philosophy*

That preceding phrase sums up Mr. Heisenberg's greatest scientific contribution.

In 1927, with its adoption of the Heisenberg uncertainty principle, quantum mechanics made it clear (at least as far as they were concerned) that indeterminism and uncertainty were essential features of the physical world, cause and effect was improbable, and physical predictions could be made only in terms of mathematical probabilities. In the scientific texts, this is reduced to the "law" that something's position and momentum cannot both be precisely known.

Figure 86 Uncertainties. Photo by S. Geng.

There is an element of uncertainty in most things. Admitting it is called honesty. But how warranted is it to place your trust in someone who puts uncertainty on a pedestal? Would you entrust your safety on a flight from New York to London to someone who gloated about how uncertain he was regarding the controls and software that guided the plane? Do you hire the guy to build a bridge across your river who puts at the top of his resumé his uncertainty concerning the strengths of structural materials and best construction practices? This uncertainty is not science. This is Voodoo. Many physicists will tell you that with the possible exception of a few things at a microscopic level, the uncertainty principle has very little use.

To gain a better understanding of the principle of uncertainty, it is illuminating to consider its opposite. Ask a quantum mechanic what is the opposite of uncertainty and if there is a pencil and paper available he should give you something that looks like this:

$$\left(\frac{1}{\triangle x \triangle p \geq \frac{h}{4\pi}} \right)$$

Actually, although that is indeed what he might give you, that answer is complete nonsense.

The opposite of uncertainty is knowledge

This is not fluffy philosophical tomfoolery. Knowledge is not just a rote recitation of formulas and facts; *knowledge is the certainty achieved by observation, study, and experience that enables you to control something.*

When a gardener is having trouble with his fruit trees and needs to consult someone for advice, whom does he choose? Someone who is free with advice but has never seen a fruit tree, or someone who has been gardening his entire life, can instantly diagnose the problem just from the description of it, and can supply a remedy he has successfully used many times on his own trees?

The wise human has knowledge—certainty—that he is able to use to control things.

Heisenberg's uncertainty principle, as it has come down to us over the past 100 years, is the exact opposite. It may not state that "everything is uncertain," but it fosters a climate where people believe that uncertainty is knowledge, when it is the exact opposite that is true.

To succeed in the pursuit of Energy Miracles, one needs to increase his or her knowledge and certainty of the fundamentals of energy as outlined in this book.

Introducing Werner Heisenberg

"The less we know about something, the more complicated it is;
and the more we know about it, the easier it becomes.
This is the simple truth about all complexities."

—Egon Friedell[97]

[97] *Cultural History of the Modern Age*, Knopf Publishers, New York.

The early 1920s witnessed fundamental difficulties in quantum physics. Numerous quantum models had been designed to explain and predict the atoms and molecules of the elements in the Periodic Table. By 1923 the quantum mechanical elite were growing frantic at the realization that all these models were failures, unable to even explain some of the simplest atoms and molecules. By the latter part of that decade, Neils Bohr, together with a group of young Germans, determined that the entirety of classical physics as well as most of the original quantum mechanical theories must be cast aside in favor of an entirely new system of concepts of physics, which was to be rebuilt from the ground up. They called this new theory quantum mechanics, and its original purpose was to accurately understand and describe the structure of atoms. From that modest beginning, they quickly made the extraordinary leap in logic that because all matter and energy were composed of atoms, this new science would necessarily define the structure and behavior of all energy and matter, whether small or large.

One of the key pioneers in that group was Werner Karl Heisenberg, a German theoretical physicist. He did his most important work between 1925 and 1927, and it is largely to him that we owe the formulation of the quantum mechanics. The most quoted of his theories is Heisenberg's uncertainty principle, which is at the core of the quantum mechanics. It declared basic concepts of classical physics such as "particle" and "wave" to be inapplicable. It also had a profound effect on a far more general level, in that it essentially invalidated the long-held and fundamental scientific thesis of cause and effect.

In 1933 Adolf Hitler became Chancellor of Germany. He was not an unknown quantity. He had already served a jail sentence for sedition and had made his demented, hysterical goals well known in his speeches and his broadly distributed book, *Mein Kampf.* By 1933 the Nazi paramilitary organization created to carry out Hitler's goals (the SS) was already well established, and within months of ascending to office, Hitler was busy rearming the German military in anticipation of waging war on the rest of Europe. The Concentration Camps were on the drawing board, Aryan Science was on the rise, and Nazi students were taking command of universities ensuring that non-Aryan and politically unacceptable faculty members were expunged.

Two other events occurred in that year of 1933. First, Germany's top physicist, Albert Einstein, along with the vast majority of German scientists, departed Germany, unwilling to grant any semblance of support to Hitler's Nazi regime. Upon his departure, Einstein said: "As long as I have any choice, I will only stay in a country where political liberty, tolerance, and equality of all citizens before the law prevail.... These conditions do not obtain in Germany at the present time."

And the second event, a 32-year-old Werner Heisenberg, who had just won the Nobel Prize in Physics, chose to remain in Hitler's Germany to serve as the Director of the Nazi Atomic Bomb Project during World War II. Heisenberg, for reasons best known to the wave function of Schrödinger's cat, fully expected a German military victory. But what follows is an example of another important thinker of the period whose approach to the Nazis was a little different.

Living only a few hundred kilometers from Heisenberg at the time was Egon Friedell, a prominent Austrian historian whose philosophy, science, and life were quite at odds with those of Heisenberg as can be seen from the quote at the beginning of this section. Friedel, a Jew turned Lutheran, stood up on his hind legs and openly described the Nazis as belonging to "*the realm of the antichrist. Every trace of nobility, piety, education, reason is persecuted in the most hateful and base manner by this bunch of debased menials.*" The Nazis then banned his books.

Friedell knew he was risking arrest and banishment to the newly constructed Concentration Camps outside Vienna. But he chose to remain true to his own reality and continued to speak out against the German Third Reich. On March 16, 1938, at about 22:00, two Nazi Storm Troopers (SS) arrived at Friedell's house to arrest him. While they were arguing with his housekeeper, Friedell, unwilling to cooperate with the Nazis in any way, jumped out of a window to his death with a determined smile on his lips, after warning some passersby to stand clear for their own safety.

As for Heisenberg, though never officially joining the Nazi Party, for five and a half years, he marched a perfect goose-step in support of the Nazi regime. He performed the Hitler salute before all public lectures, signed all official correspondence with "Heil Hitler," joined in pro-Nazi marches, outings, and indoctrination camps, and associated with rabid Nazi government bureaucrats. More ominously, he also led the pursuit for a German nuclear device that could win the war for his Fatherland. At the same time, he performed the duties of a goodwill ambassador for the Nazis, traveling around a Europe of conquered nations, extolling German culture, science, and morality. One of these trips bears mentioning.

His 1941 meeting with his old mentor, the Dane, Neils Bohr, has given rise to lots of speculation about the true nature and motives of Werner Heisenberg. It was in the middle of World War II and the two men had been out of touch for several years on account of the hostilities. Now, just a few months after the German occupation of Denmark, Heisenberg traveled to Copenhagen and insisted on a personal meeting with Bohr. Though they had been friends and collaborators for 20 years, the meeting was a disaster and famously marked the end of their friendship.

Michael Frayn memorialized this meeting in his quantum play, *Copenhagen*. The play metaphorically puts Heisenberg on trial and poses the question: Why did he come to Copenhagen for that meeting with Bohr? In the course of the investigation, alternative reasons are presented as to his actual motives, and as the play progresses, the playwright seeks to make it harder and harder for the audience to determine the truth. He did this purposely in line with Heisenberg's uncertainty principle itself: "The closer you look, the less you seem to know."

There are several possible reasons Heisenberg went to Copenhagen:

1. As Heisenberg would have us believe, he went because he was an honest and moral member of the Third Reich, who understood that Germany was going to win the war and was traveling there to extol and protect good old German virtue, culture, and science.
2. He went because he missed the company of his old friend, Bohr, as well as his favorite Smørrebrød (Danish open-faced sandwiches) and Kartofler (caramelized potato salad).
3. He went in order to recruit Bohr to aid him in persuading the U.S. and British physicists to delay the production of their atomic bomb, to ensure that Germany would not be destroyed by a nuclear bomb in the event she somehow did not win the war.

The problem with the play (as in most discussions of the meeting itself) is that it dissolves into too much opinion. Of course, it is totally possible to ascertain the truth, **but only if one looks harder**. For starters, it is probably all there in n-dimensional technicolor within the tens of thousands of pages of Heisenberg's personal letters and papers the Heisenberg family still refuses to release to scholars, though it has been almost 50 years since his death. If these papers exonerated him, it is safe to assume they would have been released long since. Their continued obfuscation does not shine a very pretty light on Heisenberg's elegant reputation. More importantly, the issue of why he went to Copenhagen begs the bigger question: Why did he work for Hitler's Germany for five and a half years, actively leading the Nazi Atomic Bomb project? If he was not a Nazi sympathizer as he later proclaimed, he was either a cowardly sparrow trying to save himself, or he was following his own uncertainty principle and just was not looking. Neither is very reassuring.

Quantum mechanics has done its best to minimize the significance of the preceding facts, claiming Heisenberg actually opposed Nazism and did his best to delay the German nuclear program, but there is absolutely zero documentation to substantiate either of these justifications.

What effect did Heisenberg think it would have for a newly minted Nobel Laureate, the most famous physicist left in Germany, to so slavishly associate himself with the barbaric politics of Adolph Hitler? The fact is it made a lot of difference. And if Heisenberg had possessed either courage or ethics, he would have refused to cooperate or just emigrated.

But Heisenberg was one of those who did not see causes and effects, and therein lies the great danger in embracing either his philosophy or his science.

An ode to Heisenberg

"Heisenberg alone was unconfined,
Too mad for classic logic chains to bind.
Now to pure space he lifts his manic stare,
Now, running 'round the circle, finds it square."

—Attribution is an uncertainty

Figure 87 The possibility exists that one or more of these, at least some of the time, was Werner Heisenberg. Or, maybe not.

Uncertainty: the predictable and the unpredictable

As discussed earlier, quantum physics got its start in the contemplation of particles in a turbulent and chaotic flow of liquid. There is some certainty there, for instance how a particle will respond to the addition of heat to the liquid, but also uncertainty in which way a particle might jump or how far it will jump. Same with the breaking of waves on a seashore. Ask any surfer and he will tell you a great many things that are certain about breaking waves, including that they will almost always break toward the shore, but there are also uncertainties involved, including shark attacks. Same holds true for the weather; there are lots of things that are certain about the weather on any given day, and there are many things that are not. In the past, scientists have erred in believing they knew everything. But the current crop errs in the opposite direction, believing they know nothing and everything is in chaos.

There is always a balance between what you can predict about something and what you cannot predict about it. It is someone's ability to stand up to and handle the unpredictable things that separate the scientists from the lab rats. **A useful scientific concept would be a measure of the degree of randomness: the ratio of the predictable factors to the unpredictable.** A small child, going to school for the first time, can be overwhelmed by the unpredictability of all the strange people and activities and the lack of his parents. Another small child, going to the same school, takes all that in stride and has a grand time playing with new friends. Quantum mechanics was dreamed up by a bunch of folks in the first category; they got a glimpse of a little too much randomness, and you can imagine the rest.

Figure 88 A quantum mechanic confronting confusion.

There is a darker aspect to this. That which a person works hardest on, he usually winds up having. You can see it all around you; devote energy to something and more often than not you end up getting it. Diplomacy is an example. As long as leaders are putting time, money, energy, and communication into diplomacy, you have diplomatic-type relations. At least people are talking instead of throwing bombs at each other. When someone comes along and fires the diplomats, boosts military spending, and starts threatening to press the nuclear button, the wars are just around the corner.

With uncertainty, chaos reigns. With certainty, progress can be made because science can be counted on to supply that progress. But for almost 100 years, students have been taught that certainty is near impossible. For close to a century, students have been taught that duplication is impossible, what you see is not really what you see, and the more you communicate with something the less you can know about it.

The more energy and credibility you put into uncertainty, the more uncertainty and chaos you will have. Welcome to the 21st century.

Measurement and Uncertainty

"One accurate measurement is worth a thousand expert opinions."

—Albert Einstein

"To measure is to know."

—William Thomson (Lord Kelvin)

Measurement is a system of refined observation the purpose of which is to help identify something or compare one object or event with others. You resort to measurement if you need more accuracy than would be afforded by a quick look or listen or smell or feel. That could mean taking out a tape measure or checking your watch, or for scientific purposes, it could mean the use of sophisticated equipment. Measurement systems have been the cornerstone of commerce and science for at least 6000 years: He runs faster than she runs. That load of vegetables is heavier than this one. It will take 5 days to get to the water hole. The British transistor responds quicker than the French one. Measurement.

There have been many systems of measurement developed since the one in ancient Egypt described in the Bible. The metric system, first adopted by France in 1799, is the most widely used today although a few countries (the U.S. and China among them) continue using some of their old units.

Measurement is one of the quantum mechanics' most insidious misrepresentations. They call it their "measurement problem" and explain it by saying that it may be all well and good to take accurate measurements in the real world, but in the quantum world the very act of observation or measurement irrevocably disturbs the results, so measurement is impossible. That is meant to persuade us that the best we can hope for from science is a "probability prediction obtainable only from quantum theoretical mathematics." When you hear that, realize you are being deceived by a patchwork of smoke and mirrors.

In February 1927, Heisenberg turned his attention to uncertainty. He had been musing about the fundamental quantum properties of a light particle (photon) and the electron in its orbit around a nucleus. How could they be measured? One day, he imagined using a gamma ray microscope to study the motion of that electron, but it occurred to him that the very act of measuring the electron's properties by shining gamma rays on it would disturb the electron in its orbit. He was correct about that, but from there he jumped to the amazing conclusion that the very act of observing any particle would alter its behavior and thus "nullify the objectivity of the observation." He then concocted a mathematical formula "proving" that mere observation of a particle prevented its location and velocity to be simultaneously measured.

He should have paid more attention in his laboratory courses. First of all, in 1927 there was no such thing as a "gamma ray microscope." In fact, at this writing, 94 years later, there still is no such microscope. He might as well have said, "We'd better cease research into heart disease because shooting a bullet through a person's heart will interfere with the heart's behavior and nullify the objectivity of the observation." Of course, you cannot use a bullet to inspect a heart. The challenge in measurement lies in devising methods of observation that do not destroy the item under test and will give repeatable results. This has always been true of all measurements, large and small. Heisenberg just was not up to that challenge.

Using a flashlight together with a simple magnifying glass and precision calipers to determine the width of a dead spider's legs is unlikely to significantly mess up the measurement.

Compare that to the actual procedure quantum mechanics use today to visualize a defenseless fermion.[98] They first reduce its temperature and, therefore, its movement, low enough to image it. A point just a few ten-thousandths of a degree above absolute zero (−273°C) is not adequate, so the "researchers" have to cool the fermion down even more. They do that by creating an optical lattice, using laser beams to form an arrangement of light "wells," which magnetically trap and hold a single fermion in place. Once "caught," the researchers put the poor fermion through a number of stages of laser temperature reduction and evaporative cooling of the potassium gas used to encase the fermion, to reduce the temperature to just above absolute zero. This is enough to hold individual fermions in place in the optical lattice. At that point, so the

[98]Fermion is a category of quantum mechanics invented particle, which they define as "very small and has an odd half-integer spin."

theory goes, the fermions have been put into such a low energy state that they are forced to release photons of light that can be captured by the microscope and used to locate the fermion's exact position within the lattice.

Such treatment should be in violation of the United Nations Convention Against Torture. You have taken a fermion, frozen it, shot it repeatedly with lasers, surrounded it with poisonous gas, and trapped it in a magnetic light well. Can these researchers really think that this utterly invasive and erosive procedure is going to have no effect on whatever it is they are calling a fermion? Whatever it is they think they are measuring by the end of this process will bear no relation at all to whatever it was before they began it. And that is a procedure being done today by researchers in the name and under the banner of Werner Heisenberg. Can you imagine anything more ironic?

Heisenberg, facing a problem of how to observe and measure a photon had several choices:

1. Use a procedure to observe the photon that would destroy or badly alter it.
2. Invent a procedure to observe and measure the photon that would neither destroy it nor alter it too much.
3. Throw in the towel and abandon measurement.

In the early days of the 20th century, it may not have been possible to detect a photon without destroying it, but instead of working a little harder to come up with better methods of observation or measurement, Heisenberg concocted a theory that it was impossible to **ever** do it. In recent years, he has been proven wrong on this, of course, since physicists are now able to detect individual photons without destroying them. Because he was inexpert in conducting real laboratory experiments, Heisenberg could not conceive of the preceding second method even though that was the only real solution. Instead, he used method one to justify method three. Bad choice.

If you are thinking the preceding situation may not be such a big deal, we will elucidate this point further with a comparison between the classical and quantum mechanical descriptions of how to observe and measure a simple event: A girl named Sylvie is tasked with going down to the river to fetch water.

Figure 89 Bring me a little water, Sylvie.

The classical description of the event follows:

Mother tells Sylvie to fetch water. Sylvie picks up the water bucket, walks down to the river, fills the bucket, returns to the house, and hands the filled bucket to mother who checks her watch and notes 3 hours have elapsed.

What follows is a very abbreviated description of the same event, using quantum measurement based on Heisenberg's uncertainty principle. The quantum version starts off with an impossible demand: that "every possible scenario" be considered.

Mother tells Sylvie to fetch water. At the same time, Sylvie has no mother. Sylvie picks up an aqua-blue water bucket with her left hand. She also picks it up with her right hand, and the bucket is sometimes red and sometimes green. On the way to the river, she is eaten by a wolf, which takes 5 min or 3 hours; accosted by perverts, which takes 3 hours or 3 days; gets tired and takes a nap lasting 1 hour or all day; meets her boyfriend for a tryst; stops at a friend's house for a snack, which is quick or long when she must first fry the chicken; gets lost for 2 hours or 2 days; and arrives at the river in 1 hour, or three, or eight or the next morning, or sometime the following week. All these events are happening simultaneously according to quantum measurement. When she gets to the river, she fills her bucket with water and half-fills it and does not fill her bucket with water at all, but substitutes mud. We have seen the bucket in her left hand. but it is also in her right hand, and her return trip from the river is filled with as much complexity and drama as the first leg of her journey. Now we come to the interesting part. According to quantum mechanics, all these things are going on at the same time, and predicated on the assumption that no one makes any attempt to observe or measure the action, because any such attempt will irrevocably ruin the event. Therefore, when she gets home and her mother **sees** her, the entire house of cards described in this paragraph collapses.[99] Sylvie is suddenly back home with a full bucket of water, which she gives to her mother, who checks her watch and finds 3 hours have elapsed. (And, surprise, no sign of wolf-bite marks.)

Of course, we do not want our Sylvie to get into trouble on the way to the river, and if she did, we would better find out about it so she can be rescued and someone else can go for the water. But the point of the above is that the quantum approach is incapable of providing any help in resolving actual chaos in real situations due to its built-in complexity and misconceptions. This holds true on both macro- and microscales.

Returning to measurement, the quantum mechanics insist that the exactitude of measurement required at the quantum level makes it completely different in character from that done in the macro world and the quantums have the monopoly on the solution to this problem. Is that true?

How precise is precise? How accurate is accurate? The uncertainty of instrumentation results and the variation of ambient conditions in measurement have been appreciated for a very long time. There is an entire industry devoted to it called Calibration. The challenge in many international standards is to find a way to measure the extent of something without influencing that something too much, and to do it in a repeatable manner so the results can be duplicated and verified. This is true whether you are measuring the distance to the moon or the distance between electrons.

There is a law about this. **You measure with the precision necessary to accomplish a specific purpose. If you cannot measure directly, you devise a workaround. And if you cannot figure out a workaround, you give it to a better scientist.**

We have set the stage for the next section, which deals with one of the most famous quantum conundrums, Schrödinger's Dead and Alive Cat, devised by Erwin Schrödinger to expose the idiocy of quantum measurement.

[99]The quantum mechanics describe this as "the waveform collapsing."

Schrödinger's Uncertain Cat

"I know an old lady who swallowed a cat.
Imagine that. She swallowed the cat.
I don't know why...
I guess she'll die."

—Rose Bonne

The physicist Erwin Schrödinger introduced his Cat Paradox in 1935. She is probably the most widely swallowed cat in history as well as the most indigestible. It was not at first a **CAT**astrophe, because when Schrödinger conceived it, it was only to illustrate the illogic in Heisenberg's uncertainty theory. (Schrödinger called Heisenberg's mathematics in defense of the uncertainty principle: "Repulsive.")

The paradox goes like this: A cat is placed in a sealed box together with a bottle of poison and a decaying radioactive substance that at some point will activate a Geiger counter connected to a trigger mechanism set to open the bottle of poison and kill the cat.

Figure 90 Schrödinger devised his famous Cat Paradox to spotlight the idiocy of quantum mechanics.

The question is, to someone outside the box: "Is the cat dead or alive." Per Heisenberg's uncertainty principle, if you put the cat in the box and there is no way of knowing whether the poison has been triggered, you have to treat the cat as if it is doing all possible things at the same time, that is, being dead **and** being alive. If you try and predict the status of the cat, you will have a 50% chance of being wrong. But if you assume it is a combination of all possible states, then you will always be right. So, the cat is both dead and alive. At least so says Herr Heisenberg and the quantum mechanics.

Schrödinger rejected that interpretation and dreamed up the preceding "thought experiment"[100] to expose its idiocy. The weirdest part of the story is that when Heisenberg heard about it, he thought it was the **PURR**fect way to illustrate his uncertainty principle and it remains just compli**CAT**ed enough to suck people into the mystery of it all. Since then, many people have swallowed this sophisti**CAT**ed cat.

To his credit, Schrödinger's contribution does highlight one of the basic weaknesses of quantum mechanics.

Heisenberg accepts the premise "until the box is opened, an observer doesn't know whether the cat is alive or dead," but this is far from a defensible assumption. He could knock on the door of the box and see if there is a response. He could listen for any activity in the box. He could install a security camera inside the box or just phone the cat's cell phone and ask her for an update. Or if he were in really good shape, he could just "know." How many times have a pair of identical twins, or husbands and wives, or good friends, though separated by hundreds or thousands of miles suddenly become aware of the death or emotional trauma of their loved one? Of course, we would not expect Heisenberg to be capable of such an affinity.

The Cat Paradox shows that instead of doing some honest work and devising new methods by which to observe the cat, Heisenberg just cops out with the assumption "there's no way of observing the cat."

The concept of uncertainty is not at all new. Irish Philosopher George Berkeley was tossing around his own version of the uncertainty principle 200 years before Heisenberg: *If a tree falls in the forest and nobody is there to hear it, did it make a sound?*

The solution, by the way, if you really need to find out whether the tree made a sound, is to send someone into the forest to observe and to listen.

[100]See sections on "The Observer" and "Thought Experiments" in this book.

Clarification by the cat, herself

(*A note discovered among the private papers of Erwin Schrödinger*)

Hi. I am Schrödinger's cat—frequently blamed for the rapid spread of quantum physics because I was supposed to have been in a box both dead and alive. At the same time, yet! What a bunch of dog poop!

Yes, I admit I was in the box. And I admit that the box was sealed. And back in 1930, it might have been a little hard to tell what was going on in the box. But "dead and alive at the same time?" Give me a break!

Erwin, my owner, invented the story to demonstrate the absurdity of quantum mechanics. He called it a "ridiculous case." But two or three guys (not the brightest pups in the litter) decided it proved I indeed **was** both dead and alive. The same geniuses went on to claim I was proof that you could never know anything with certainty. But to my mind, all they really proved was that they were a bunch of lazy dogs. There are 100 ways they could have determined what was going on in that box:

They could have measured the heat of the room where the box was. If I had died, the temperature would have dropped. They could have measured the weight of the box with me in it. (They probably did not even know that at the moment a cat dies, she instantly loses an ounce or two.) They could have measured the sound waves in the walls of the box. Those waves would have changed if my breathing had stopped.

In 100 years, they will doubtlessly come up with fancy infrared cameras able to instantly and accurately detect the presence of live cats in boxes.

And more to the point, they could have just asked me. Has not anyone ever heard of animal rights? They could have knocked on the door, and I could have told them.

I do not like my name being associated with all that quasi-scientific canine manure, and I certainly would not like my grandchildren to think I had anything to do with it.

That is why I leave this note behind.

Attested as true.
Schrödinger's Cat

Standard Model of Particle Physics

"The atoms or elementary particles themselves are not real;
they form a world of potentialities or possibilities
rather than one of things or facts."

—Werner Heisenberg

Good physics has always been simple. The great discoveries have always been related to simplification. Examples are Newton's law of gravity. When science believed that all the planets rotated around the Earth, there were countless "natural constants" needed to keep the books straight. The motion of every planet required two of these "constants": one called its deferent, and the other its epicycle. These were just fudge factors[101] and could all be instantly eliminated once Newton's law had been rolled out. Maxwell's electrodynamics unified electric and magnetic constants into one formula, halving the factors you needed to worry about.

Here's Niels Bohr's original model of the hydrogen atom. Not perfect, and not workable in all cases, but very useful and very simple.

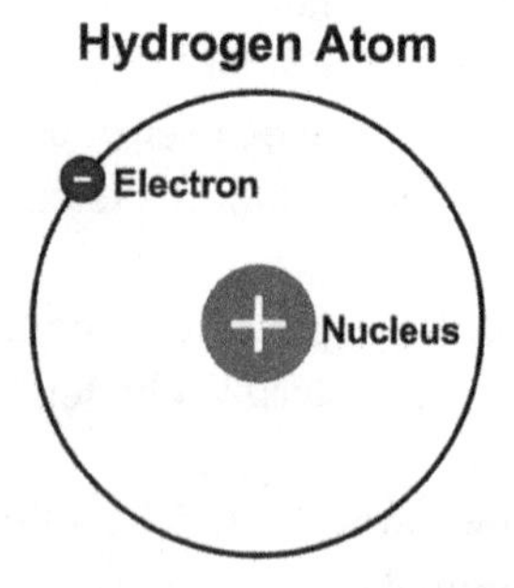

Figure 91 Neils Bohr's original model of a hydrogen atom: a negatively charged electron moving around a positively charged nucleus.

Compare that with today's quantum mechanics "standard model" of particle physics, with its huge number of theoretical particles with adjustable parameters, which grows ever larger as more particles and characteristics are invented.

[101]A "fudge factor" is a quantity or element introduced into a calculation or formula in order to make it fit the scientist's expectations. It is generally a solution to a specific problem, not applicable to broader purposes.

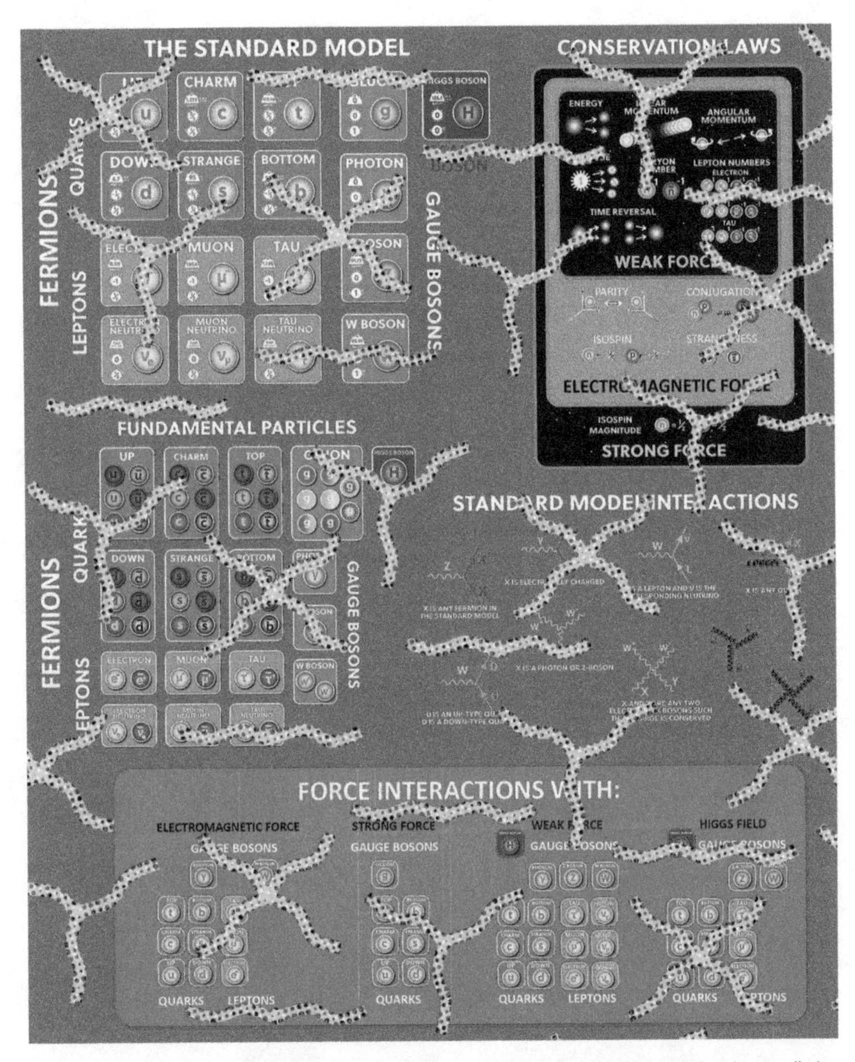

Figure 92 Above, in the background, is the "standard model of particle physics," the quantum mechanics attempt at describing how the universe is put together. Despite (rather, because of) its dozens of invented particles and attributions, this theory has so many deficiencies that the quantum mechanics themselves are now feeling the need to dispense with all those imaginary particles and replace them with their new theory (shown in the foreground) called "String Theory." There is no need to try to explain String Theory here since a picture of a gaggle of strings is worth a thousand words.

There are no actual grounds for any of these particles—no empirical evidence to support them—and they are entirely contrived from the imaginations of quantum mechanics. Figure 92 is a recent diagram of the modern and simplified "standard model of particle physics."

Not only can you not accuse it of being simple, but it may not be any more useful than the original Bohr model on which it is based. According to an international team of chemists (including the Nuclear Materials Processing Group, Oak Ridge National Laboratory, the Laboratorie de Physique et Chimie des Nano-objets at the French Institut National des Sciences Appliquées, and the German Institut für Anorganische Chemie), the preceding "standard model of quantum mechanics" with its dozens of carefully crafted particles and invented characteristics still fails to explain the behavior of almost 20% of the elements in the Periodic Table.[102]

It may not be inevitable, but it certainly appears that the great discoveries in physics are simple ones.

P.S. You will be glad to know that quantum mechanics have been able to furnish us with an analysis of the secret workings going on deep inside the atoms of your brain as you have been reading this. If you were particularly perceptive, you might have sensed the annihilation of the electron and positron that just occurred heralding the celebratory Birth of the Blue Proton (ϒ). Did you feel it? If you did not, do not worry. Figure 93 will make everything clear.

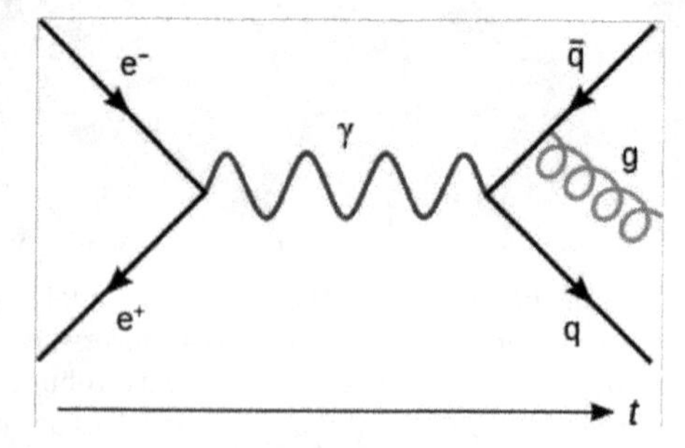

Figure 93 Feynman diagram showing the collision of imaginary particles. There are "laws" that govern them, but because they are only "virtual" particles that come into existence for a mere ghostly instant, they are expected to violate all of those laws.

---

[102]Periodic Table's heaviest elements alter theory of quantum mechanics. *Science News*, October 4, 2017. (Research published in the *Journal of the American Chemical Society*).

The key, you see, is the gluon. Once "born" the proton (Υ) morphs into a quark–anti-quark pair, which then inexplicably radiates the Green Gluon (g). All these so-called particles are imaginary, but even so could not be made to balance out mathematically no matter how many times the pocket calculator was kicked.

To "fix" it, quantum mechanics labels the positron an "antiparticle" and sends it backward in time, shown in the diagram as e^+ moving from right to left. This is the same diagram quantum mechanics use to explain "retro causality," the concept that an effect actually precedes its cause in time and so a later event can affect an earlier one. I am not making this up. This is 21st-century quantum science.

To be clear, the major difference between me and the quantum mechanics is that I am at least honest enough to admit I have no idea what an atom is composed of.

TOE (Theory of Everything) or TON (Theory of Nothing)?

"I have become a lonely old chap who is mainly known because he doesn't wear socks, and who is exhibited as a curiosity on special occasions."

—Albert Einstein shortly before his death

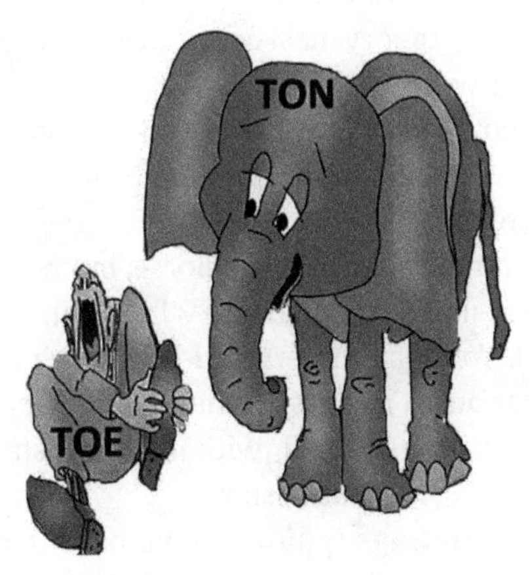

Figure 94 TOE (theory of everything) or TON (theory of nothing)?

Einstein's 30-year quest had ended in failure. The more he thought about it, the more he realized that his theories fell far short of being the complete explanation of the universe that had been advertised. In particular, it pained him to admit that his General Theory of Relativity failed to explain either gravity or electromagnetic waves. The emergence of the quantum mechanics, unabashedly based on his relativity theory, pained him even further.

It was patently clear to him that relativity and quantum mechanics could not both be correct. In that he was only partially right. While true that both could not be correct, he missed the third (and ultimately correct alternative) that **neither** was correct. Consequently, his decades-long search for a unified theory ended in bitter disappointment. The search did not end with him, however. It seemed a worthy quest for lots of theorists and mathematicians with not enough wit but too much time on their hands.

They called it the Theory of Everything, and they defined it as, "the final theory, ultimate theory, or master theory; a hypothetical single, all-encompassing, coherent theoretical framework of physics that fully explains and links together all physical aspects of the universe."[103]

The quantum mechanics' latest iteration of the Theory of Everything is the "string theory" that tells us that all the currently imagined "basic particles" like quarks and neutrinos are themselves "actually" made up of even more exotically imagined string-like elements. The string theory has never predicted anything about anything and never will.

There is not now, nor there has ever been one understanding that explains all. As most people would appreciate, there are *many* understandings (plural).

You may understand how to ride a horse, but not how to balance the books. You may understand how to wire a home, but not how to weld sheet metal. You may understand how to be a great CEO but not how to be a spouse. You may understand every component of a five-star gourmet French meal, without understanding the first thing about controlling nuclear fission.

So, it is **understandings** (plural). Remember that the next time you hear the quantum mechanics, who have given us so little of use

[103]Steven Weinberg, *Dreams of a Final Theory: The Scientist's Search for the Ultimate Laws of Nature*, Knopf Doubleday Publishing Group.

in the last hundred years, telling us they are close to finding the single understanding that explains it all.

Spacetime and Fourth Dimension

"There was a young lady named Bright,
Whose speed was far faster than light.
She went out one day
In a relative way,
And returned on the previous night."

—Arthur Henry Reginald Buller

Stephen Hawking explains spacetime as follows:

> *"General Relativity is a theory not only of curved space, but of curved time as well. Einstein had realized in 1905, that space and time are intimately connected with each other. One can describe the location of an event by four numbers. Three numbers describe the position of the event. They could be miles north and east of Oxford circus, and height above sea level. On a larger scale, they could be galactic latitude and longitude, and distance from the center of the galaxy. The fourth number, is the time of the event. Thus, one can think of space and time together, as a four-dimensional entity, called Spacetime."*[104]

Whereas space and time may well be intimately connected to each other, they are not *that* intimately connected. And the preceding explanation directly defies Einstein's relativity.

Einstein was the first to realize the relative nature of time. That is to say, there is no "absolute" system of time; no "big clock up in the sky" that serves as the reference point for all Earth clocks. There are many systems of time, and clocks have little or nothing to do with most of them. You can say, "June 14, 1946" (which refers to a system of time in use only on this planet) or "the morning of the birthday of the mule" or "3 years after he graduated" or "just before the accident" or "five generations ago" or "when I was young and wild" or "right after the storm." These are all equally valid systems of time and as can be seen, they have little or nothing to do with either clocks or a location in space. Time is not a ticking clock, and clocks create

---

[104]"Space and Time Warps," a lecture by Stephen Hawking.

neither space nor time. Time marks the co-action of particles or the change in position of particles in space. The common denominator of time is change, and change is its primary manifestation.[105]

What Einstein failed to do was take this one step further. If time was indeed a relative characteristic, then location was equally so, and there are as many different systems of pinpointing location as there are of time. An observer could just as validly designate a location as "under the tree" or "midway between the two mountains" or "where the hanging occurred" or "5 miles outside of town" or "some place on planet Mars" or "10 light years away."

Every quantum mechanics adherent is trying to insist on his own particularly favored system of time or location. Most are partial to years, months, days. Others prefer latitude, longitude, distance. But in doing so, they are all violating the relativity principle. The preceding is mentioned to introduce the more important factor here: **dimensions**.

n-dimensional space

Quantum mechanics seem to get their kicks out of dreaming up and then disseminating mystic theories such as the fourth dimension. Actually, they do not leave it there. They even imagine "n-dimensional space" referring to a space with unlimited dimensions.

The dictionary defines dimension as "a measurable extent of some kind such as length, breadth, height, etc." Note the word **measurable**. Classical physics observed that the physical universe existed in a space of three dimensions.

Quantum mechanics have no definition for space. A mathematician will talk about "points." He or she will say a point is something with location but without dimension. It has no length, breadth, or thickness. One point may exist in mathematics, but not in physics. You can have a single geometric point, but you would need at least two points in order to have a "measurable extent of something." With one point, there can be no space since space is a viewpoint of dimension, and you would need a minimum of two points to satisfy that. Two points give you a linear (one-dimensional) space.

---

[105]See section on Time in this book for additional information.

Figure 95 One-dimensional space requires two points.

Physics has taught us about three-dimensional spaces. Those observations are valid, and they cover all the spaces in this universe. The eight corners of a cube are an example of three-dimensional space.

Figure 96 Three-dimensional space.

Every sentient being on Earth (with the exception of a few quantum mechanics) will tell you that space and time are not identical. They may not be able to define these two terms with exactitude, but they will know with certainty that they are different. When you tell a friend "meet me after work," you both know you are referring to time. When you say "let's go to the seaside," it is clear you are referring to a location in space. Einstein, finding himself backed into a theoretical corner with his legacy in jeopardy, was forced to advance the notion that space and time were identical. Space equals time equals space. You could not have one without the other, and so he joined them at the hip and explained the imaginary elastic property he imputed to it: "If you stood on it, things nearby would roll towards you, like a trampoline." Figure 97 shows a red ball stretching a trampoline in a delusional four-dimensional space.

According to Einstein, the action of energy and objects would cause his trampoline, or spacetime, to "curve." And that curvature is the quantum mechanic's "fourth dimension." It makes for an awesome visual graphic, but as can be seen in Fig. 98, it fits very handily into a simple three-dimensional space.[106]

[106]See sections in this book on Time and Space for more on this topic.

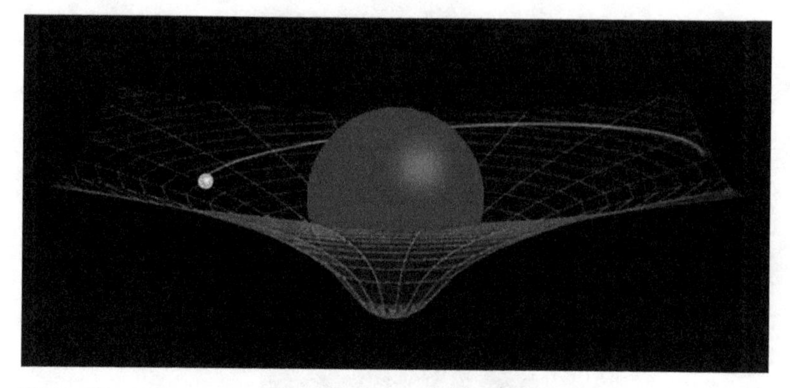

Figure 97 Delusional red ball stretching delusional spacetime.

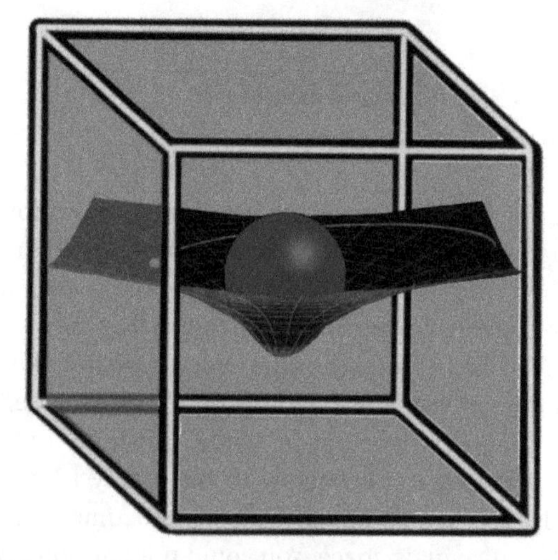

Figure 98 Delusional spacetime fitting nicely into three-dimensional space.

Einstein rejected the concept of an aether because its existence would have invalidated his special and general theories of relativity. In its place, he substituted a trampoline he called "spacetime."

Space and time are not identical. And the quantum mechanic's theoretical "spacetime" does not give us any additional "fourth dimension." Some will find that disappointing, but it is nonetheless factual. An appreciation of the preceding fact will put one on the road to locating some Energy Miracles.

Figure 99 Albert Einstein on his trampoline.

Quantum Mechanics' Uncertain Zero

Once upon a time, there was a little man named Slim. He was sitting on his porch one day minding his own business and engaging in his daily 1-hour think period.

Figure 100 Quantum mechanics failure to define zero.

Slim was a quantum mechanic, and he had been deeply worried ever since the day, while working on a complex mathematical formula, he realized that he did not really understand the definition of zero. Sounds pretty theoretical, but actually it was a big deal because for the first time he realized the real origin of the quantum mechanics, and the realization had him paralyzed with fear.

Why did quantum mechanics *really* come about? What prompted Einstein to go so far astray with his relativity theories and Heisenberg to throw in the towel with his uncertainty principle?

Slim realized the answer to all those questions was the same. It was their inability to define zero. From its inception, the subject of engineering has always dealt with kinetics and statics: things that move and things that do not. But quantum mechanics had failed to define static.

To be fair, classical physics had also failed to define it. But when the quantum mechanics started trying to mess around with atomic particles, the lack of a correct definition for zero really screwed the pooch.[107] It was for only that reason that quantum mechanics had to say that classical physics was one thing playing by one set of rules, and atomic and molecular phenomena were a different thing playing by a different set of rules. That was not the case 100 years ago, and it is still not the case.

Well, what is zero? A zero condition or thing is one that has no wavelength, no physical universe location, no mass, no size, no measurements of any kind, no particles moving or not moving. This is a much more complete definition of zero than quantum mechanics ever conceptualized.

Quantum mechanics fell off the wagon by defining a static (zero) as a **particle at rest**.

Then they did not define "rest."

There are very few "absolutes" in this universe. But zero is one of them.

Zero (or a static) has no mass, no motion, no wavelength, no location in space or in time.

If, as quantum mechanics assert, a photon is a static particle with no mass, then it has no location in space or time and that is obviously not the case.

There was no need for the quantum mechanics to throw away Newton and Maxwell and the rest of physics. Everything that is true in classical physics is equally true in the quantum world of atoms and

[107]Screw the pooch: meaning "to commit an embarrassing mistake." Comes out of the NASA Mercury space program.

particles. But it is only true when you have an accurate definition for zero.

A photon is a particle of energy, not a static. By confusing energy with zero, by denying energy a location in space, by denying energy a mechanism by which it can propagate through space, quantum mechanics made it impossible for progress to be made in the subject of new energy discovery.

The return to workable mechanisms is the road to the Energy Miracles.

Quantum Favorite Sport: Shoot the Observer

> *"It seems sensible to discard all hope of observing hitherto unobservable quantities... Instead it seems more reasonable to try to establish a theoretic quantum mechanics."*
>
> —Werner Heisenberg

Figure 101 Shooting the observer (the quantum mechanics favorite sport).

The observer says: "*We know a thing or two because we have observed a thing or two.*"

Quantum mechanic response: "*The Observer is ruining the science.*"

A cornerstone of quantum mechanics is the principle that you cannot directly observe or measure an object without adversely affecting the measurement itself. This came from the inability of quantum scientists in the early days of the 20th century to detect a photon without destroying it. As shown earlier, instead of working a little harder to devise a better way to observe those photons (which was accomplished by later scientists), the early quantum mechanics decided that it was impossible to ever directly observe them, and the better idea would be to rid themselves of observation and the observer.

In fact, it is the lack of an observer that is ruining the science. Quantum mechanics and its offshoots teach you the hopelessness of looking because looking will ruin the experiment, when the truth is that only by looking at and communicating with something can you ever gain any perception or understanding of it. By invalidating the observer, quantum mechanics is invalidating the scientist.

The biggest hoax ever foisted on science was to get scientists believing it is somehow cheating to look, when in fact looking is the only way to advance.

Real Experiments versus Thought Experiments

"The principal of science, almost its definition, is the following:
The test of all knowledge is experiment.
Experiment is the sole judge of scientific truth."

—Richard Feynman

"Today's scientists have substituted mathematics for experiments,
and they wander off through equation after equation,
and eventually build a structure which has no relation to reality."

—Nikola Tesla

Science and experiments

An **experiment** is a procedure carried out to support, refute, or validate a hypothesis. Experiments provide insight into cause and effect by demonstrating what outcome occurs when a particular factor is manipulated under controlled conditions. It comes from the Latin *experiri*, "to try." Experiments always rely on repeatable procedures and logical analysis of the results. As a primary component of the scientific method in engineering and the physical sciences, experiments have always been empirical processes, meaning they utilize direct or indirect observation and experience to gain knowledge.

It has been quite correctly pointed out that progress in science requires the interplay of experimental studies and theory. Without humans advancing theories, science would be a dead-end street. By the same token, without observation and experiments being conducted to confirm the truth or workability of a theory, science can go badly adrift. In the absence of observation or actual experiment, all you have is theories about theories and this trend adversely affects many areas.

Take a police detective. In his 20 years of experience, he has personally observed many hundreds of crime scenes. He comes up with a theory that says: "Most murders are committed by family members." That may be useful or may not be useful. It is a **theory** based on his limited observation. A statistician offers to improve that theory "because it failed to calculate the standard deviation." You see, the statistician has a theory about the theory, so he goes ahead and gets the mean by averaging the number of murders for each different category. Then for each number, he would subtract the mean and square the result, then work out the mean of those squared differences, and finally take the square root of that. Wow. He would have the standard deviation; but would he have improved the theory? No, not even a tiny bit. To improve that theory would have required more and closer observation, which if done would have disclosed that the true percentage of murders committed by family members is less than 20%.

Quantum mechanics is a study of theories concerning theories about theories. It did not actually start that way. At the end of the 19th century, Max Planck was studying actual energy. It was observable that when certain metals such as iron were heated up, they would at first radiate mainly red light, then orange, then white, and finally blue. Planck was attempting, unsuccessfully, to devise an equation to describe the curve of wavelengths at each temperature. When he could not do it by direct measurement, he resorted to a theory based on statistical mathematics, but still tied to his measurements. His theory may have been more or less accurate, but it was based on observation.

Then, at the beginning of the 20th century, quantum mechanics took the simple but powerful mechanism of experimentation to an abstract cavern where theretofore no scientist worth his tenure had ever deigned to creep. Experimentation mattered so little that scientists no longer even bothered to try. It was enough to explain theories with new theories supported by mathematical formulas.

Einstein came along and advanced theories on Planck's theory. Then Heisenberg came along and advanced further theories on Einstein's theories. Then Dirac theorized on Heisenberg's theories... and so it went.

But it all started with Einstein.

Thought experiments: abstractions about abstractions

"There's a condition worse than being unable to see, and that is imagining one sees."

—Ron Hubbard

The term "thought experiment" comes from the German word *Gedankenexperiment*. The scientist Ernst Mach used it in 1883 to describe the imaginary conduct of a **real** experiment. Mach would imagine the hypothetical result of an actual experiment and then give the job to his students to test out his hypothesis by actually performing the experiment. The English phrase "thought experiment" appeared for the first time in 1897 in a translation of Mach's works.

It should be clear that Mach's use of the term exactly followed the classical definition given earlier, i.e., he actually had the experiments performed in order to prove his theories. Throughout history, people have posed hypothetical questions to themselves and others in an effort to figure things out. This is a perfectly natural form of thinking, and every famous scientific thinker from Galileo to Aristotle to Newton to Maxwell has employed the mechanism.

It was Albert Einstein who was the catalyst for the acceptance of a new, peculiar, and what was to become revolutionary view of experiment. In dozens of books and hundreds of Internet pages about Einstein, the birth of his "thought experiments" is said to have occurred in a Swiss middle school when Einstein was 16 years old and being taught according to the precepts of a Swiss educational reformer of the early 19th century named Johann Heinrich Pestalozzi. Pestalozzi believed, among other things, that he could prevent children from being alienated from the educational process by making them finance their own education through working for him. These days it would be called child labor, and his experiment in education ended badly when his Neuhof School went bankrupt after the children and their parents rebelled.

Because of his alleged influence on Einstein's development, Pestalozzi has been occasionally called the Father of Modern Education. The myth is that Pestalozzi's methods fostered Einstein's process of solving problems by "visualizing images" and engaging in "thought experiments." This had a circular effect in that Einstein's quirks in this regard could then be normalized by pointing out, "he learned it in school from the Father of Modern Education."

Only he did not. Pestalozzi preached no such thing. Pestalozzi's routine was a simple one, basically a two-step operation that started with (1) the student's careful perception of an actual something in the environment, be it an object, sound, or sight, and then (2) the student creating a visualization of that thing in his mind's eye. Pestalozzi's idea was that a student could achieve better clarity of an observed physical thing by duplicating it with a personally constructed mental image of it. Figure 102 shows the right way to learn according to Pestalozzi.

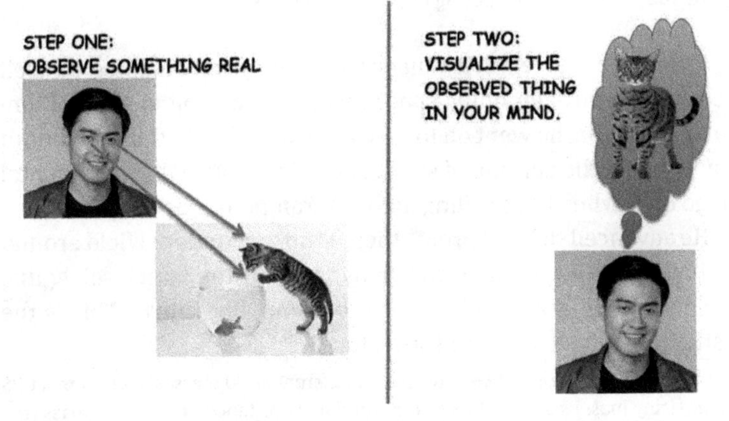

Figure 102 Pestalozzi's method began with observing something real.

Whether or not this approach is an effective learning aid is beside the point. The fact is that this is **not** what Einstein did. Einstein's "thought experiments" did not start from the observation or perception of real things. Instead, Einstein made abstractions from abstractions, corrupting the Pestalozzi process by omitting the first step. Not being based on anything factual, the "results" obtained from such action can get pretty nutty.

The thought experiment that led to his theory of special relativity stemmed from a daydream he had in middle school, when he tried to picture in his mind what it would be like to ride alongside a light beam. Of course, he had never run after any light beam traveling at the speed of light. So he was just theorizing about a theory. And it had nothing to do with Mr. Pestalozzi.

Figure 103 Einstein abstracting about abstractions.

In his general relativity theory, Einstein first imagined himself accelerating through outer space inside a speeding elevator. From that abstraction, he went on to imagine a new abstraction, this one of gravity being the curving of space and time, explainable by a mental image of bowling balls rolling across a trampoline. See Fig. 104.

He advanced that as "proof" there is no gravitational field around the balls, but that smaller balls move toward the larger solely due to the way the larger ball curves the trampoline fabric. This is the abstraction which he named spacetime.[108]

[108]Gravity is the relation between objects of different sized masses. A clock in a GPS satellite (tiny mass) slows in direct proportion to its distance from a large mass (the Earth). That is an example of one of today's most advanced measurement systems verifying the experiments of Sir Isaac Newton.

Figure 104 Einstein abstracting spacetime based on another abstraction.

These abstractions of abstractions became a hallmark of Einstein's career. Over the years, he would picture in his mind such things as accelerating elevators in outer space, falling painters, two-dimensional blind beetles crawling on curved branches, as well as a variety of contraptions designed to pinpoint the location and velocity of speeding electrons.

Both his relativity theories were the result of abstractions about abstractions.

Anyone can create a mental image picture. And anyone can change a mental image picture once created. It is one of humankind's greatest abilities. On the plus side, Einstein was adept at creating mental image pictures. On the minus side, he was not so good at observing actual things. Einstein's cavalier willingness to abandon Newton's concepts of space and time and reject the existence of a medium that propagated electromagnetic waves cannot be blamed on Pestalozzi. These unfortunate events resulted from Einstein's penchant for imagining abstractions about other abstractions rather than basing his science on observation and experimentation.

It is to a teenage Einstein that science owes this modern use of "thought experiment." It is traceable to that exact moment as a 16-year old when he became transfixed by a mental image picture of catching up to a light beam. Maybe the fixation came out of a past life incident where the result of his catching up with a beam was his being blown up in a horrific explosion. Whatever the reason, it began a chain of events that threw all of science on its head. By legitimizing such a lazy procedure, Einstein gave carte blanche for all of science to follow in his path.

The Observer: Lost Somewhere within the Three Universes

> *"Separation of the observer from the phenomenon to be observed is no longer possible."*
>
> **—Werner Heisenberg**

The quantum mechanics have reached a sorry state when the laboratory director cannot separate himself from the laboratory rat.

Figure 105 Lab director or lab rat?

Until the turn of the 20th century, the foundation of all physics was Newtonian mechanics, a subject that included concepts of energy and mass so critical to the search for Energy Miracles. We have discussed how fundamental laws governing energy and electricity were abandoned by the quantum mechanics. Equally important is the way they have changed the concept and function of the observer. By observer is meant the scientist, electrical engineer, or other persons we are depending on to discover new sources of alternative energy.

Before quantum mechanics, the physical universe, including energy, was seen as existing more or less independently of the observer. Heisenberg decided that the relativity theory changed all that, and that the role of the observer must be seen in a different light. Heisenberg became almost frantic on this point. In his own confused words, "Science is no longer in the position of observer of nature but rather recognizes itself as part of the interplay between man and nature. The scientific method...changes and transforms its object: the procedure can no longer keep its distance from the object."

Despite what quantum mechanics asserts, most people understand the difference between the lab director and the lab rat, if for no other reason than one sleeps in a bed and the other in a cage. For a lab director to accomplish anything in his rat research, he must be very clear on the fact that he is the director and not the rat. He may take good care of the rats, even become fond of the rats, but to be at all effective he would need to keep some distance between himself and the rats.

We are all continuously surrounded, every day of our lives, by real facts of nature. We observe the existence of stars, the sun, the moon, and the Earth. Almost nobody denies the real existence of cities, streets, houses, cats, and all the many objects that we can see and touch. We believe there is a difference between the person firing the gun and the person being shot. We can tell the difference between the judge pronouncing his sentence and the convicted felon. We are clear eyed about the difference between the girl who is in the store shopping and the dog food on the shelves. We know the person who smells the flower is different from the flower. Quantum mechanics teaches that none of that is real.

Many physicists are unaware that the interpretation of modern quantum mechanics implies that energy and matter do not exist independently of the observer. Quantum mechanics teach that it is somehow the observer's action of observing that brings the observation into reality: "The cat did not die unless you saw it." When touched upon in modern physics courses, professors are quick to brush aside this uncomfortable subject as unimportant, yet it is upon this viewpoint that quantum mechanics was built. And it is precisely ideas such as this that explain why no one has come up with a better source of energy over the course of the last century.

It has always been appreciated that the role of the "observer" is critical to theories of physics (including in quantum mechanics). It was key to all of Einstein's theories as well as Heisenberg's. But no one before Hubbard had described what the observer is and what it consists of. What has been missed from all discussions of the observer is the subject of universes. A universe is defined as a whole system of created things, and there are three in number: The first of these is one's own universe. The second universe would be the material universe, which is the universe of matter, energy, space, and time, which is the common meeting ground of all of us. The third universe is actually a class of universes, which could be called "the other person's universe," for all the "other persons" have universes of their own.

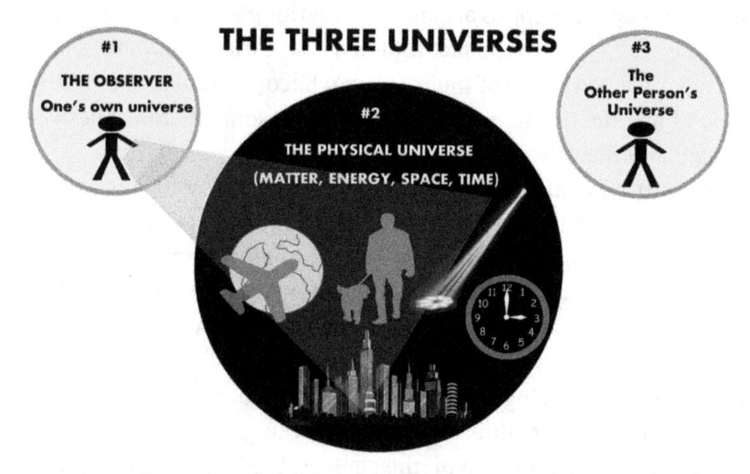

Figure 106 The three universes: (1) the observer's universe, (2) the physical universe, and (3) the other person's universe. See Hubbard, R., COHA, p.188.

Most of us educated within the last century, including the physicist, came to believe the mathematical equation, and not observation, was the most fundamental characteristic of physics. At the same time, the observer became one of the key complexities of quantum mechanics. The observer is constantly being confused with the thing being observed. It simplifies many things when we recognize that the observer has its own universe and is not a part of the physical universe. Many of the "unsolvable" thought experiments containing observers become comprehensible after unpacking the three universes.

Number 1 is the observer's own universe, quite distinct from the physical universe.

But it is in this second universe, the physical universe, where we see the mechanical effects of matter and energy in space and time, and which we are addressing in this book. This second universe, the physical universe, contains all the phenomena and laws embraced by both classical physics and quantum mechanics. It is the universe of motion and kinetics, of atoms and electromagnetic radiation of all types, including light, and is where the Energy Miracles lay hidden.

Universe 1, the universe of the observer, is separate and different from Universe 2. That is all that needs to be said here about it. Just because a lab director is capable of observing and handling a laboratory rat does not require that he be a rat. More broadly, just because the observer is capable of observing and handling matter and energy in space and time, it does not mean it is made of matter or energy or that it exists in physical universe space. The quantum mechanical requirement that only a physical entity made of energy can observe and control other physical entities made of energy is tantamount to saying that all truck drivers must be made of iron. Why? They handle trucks, and trucks are made of iron. So, by quantum mechanical logic, the drivers must also themselves be made of iron.

No. There is no place on Earth where something made out of iron can only be handled by something made out of iron. Your wife or girlfriend can open up a can of beans, though she is not made of iron. Your husband or boyfriend can use power tools to fix something, though he is not made of iron.

The truck driver who gets confused and starts believing he is the truck is going to have accidents down the road because he is no longer in control. So long as he realizes he is the truck **driver**, he can stay at cause, doing his job, and putting the truck through its paces.

This applies equally to the scientist in search of new sources of energy. Once he realizes he is not the energy he is researching, he will become more causative about his researches. Once he realizes he can observe, without having to become the thing observed, he will become more capable of handling and controlling energy. And that is where we need to be to make real progress in new energy sources.

Insanity and the Quest for an Energy Miracle

Many would have heard Albert Einstein's famous line: "*Insanity is doing the same thing over and over again and expecting different results.*" There are just two things wrong with it. One, what is described is not an accurate characterization of actual insanity. And two, Albert Einstein never said it.[109]

Stupidity is a better word to reflect the action of doing the same thing over and over again and expecting different results. And uncomfortably, this describes the previous quests for Energy Miracles, meager as they have been.

We have shown that advances in the study of electricity generation ceased in the 1930s and suggested that this followed the popularization of quantum mechanics theories concerning energy. By changing the way energy, itself, was understood and how it was propagated, these new theories effectively shut down research into new energy sources.

For over 100 years, people have taught these new theories with no results, tried to use these new theories with no results, and if they ever considered the possibility of finding a new source of energy, they based their work on these same theories, again with no results. Do we continue to repeat these same actions over and over and over, expecting a different result? We are going to have to change some things if we want different results.

While there may be nothing wrong with giving millions of computers to millions of monkeys and hoping after millions of years that one may come up with something of value, it is not science. And the human race may not have that much time.

---

[109]It is listed within a section called "Misattributed to Einstein" in the comprehensive reference "*The Ultimate Quotable Einstein*" from Princeton University Press. That exact quote traces back to a 1983 mystery novel by Rita Mae Brown "Sudden Death," where she attributes it to a fictional character named Jane Fulton.

Chapter 9

Scientific Serendipity

This chapter will examine several concepts and methods scientists and instructors were using while great electrical energy discoveries were being made, but which have since been abandoned.

When you are walking across a bridge on a bright sunny day with a spring in your step and a dry shirt on your back, you may become disoriented or confused to suddenly discover your shirt is sopping wet and you are having difficulty breathing.

When someone points out that you have fallen off the bridge and are floating in the bay, you are likely to conclude that staying on the bridge was a good formula for keeping your shirt dry and progressing across the chasm.

So it is with those searching for new sources of energy: Let us get back onto the bridge.

Mathematics Miasma

"Since the mathematicians have invaded the theory of relativity, I do not understand it myself anymore."

—Albert Einstein.

"Abandon all hope, ye who enter here."

—Dante Alighieri

Energy Miracles: The Global Warming Backup Plan
H. B. Glushakow

ISBN 978-981-4968-18-8 (Hardcover), 978-1-003-28442-0 (eBook)
www.jennystanford.com

A specter is haunting the halls of science—the specter of theoretical mathematics. Students today are taught that mathematics is how they think—it is not. They are instructed that mathematics determines logic—it does not. And most importantly in the quest for Energy·Miracles, they are indoctrinated that mathematics is the only way to communicate accurately about anything scientific—an ironic exaggeration. None of the above are true.

James Clerk Maxwell had this to say about it: "Mathematicians may flatter themselves that they possess new ideas which mere human language is as yet unable to express. Let them make the effort to express these ideas in appropriate words without the aid of symbols, and if they succeed, they will not only lay us laymen under a lasting obligation, but, we venture to say, they will find themselves very much enlightened during the process."[110]

For hundreds of years, mathematics has been a valuable tool of science and it continues to be so. But at best it is a servo-mechanism—a tool used by people—and we should take care for the tasks we set for it. Like a good shovel, mathematics can dig your grave, but it will never get you to the promised land.

In any process (computer or otherwise), it is a *person* who designs and writes the program. Most importantly, it is a *person* who interprets the results. If she is good and the program and data have been tethered to reality, the mathematics can be valuable, and there are plenty examples of this. Otherwise, you have GIGO (garbage in equals garbage out) and these days GIGO has swamped the realm of energy science.

As cited earlier, the classical definition of mathematics is:

> **"The science of quantity; the science which treats of magnitude and number, or of whatever can be measured or numbered....It is the peculiar excellence of mathematics that its principles are demonstrable."**

The emphasis is on magnitude, number, and demonstrability. Although there has always been a branch of mathematics that treated "pure," "abstract," or "speculative" math, until quantum mechanics came along, the purely abstract was always considered the baby brother of the mainline activity which dealt with the magnitudes

[110]James Clerk Maxwell, Ed. W. D. Niven (2003). *The Scientific Papers of James Clerk Maxwell*, Courier Corporation, p.328.

and relationships of things that could be sensed, measured, or numbered; i.e., real things. Measuring the heat of a volcano is the correct province of mathematics, not the heat of an imaginary devil's breath. Only the first will help those on the quest for an Energy Miracle.

In his book, *Character of Physical Law*, Richard Feynman hypothesized that "ultimately physics will not require a mathematical statement. In the end the laws will turn out to be simple."[111]

Let us take a step in that direction.

The Biggest Tiger in Energy Science: The Attack on Cause and Effect

> *"The scientist must be alert to what is* ***importantly wrong****. It is inappropriate to be concerned about mice when there are tigers abroad."*[112]

The British statistician who made that comment was observant. There are plenty of tigers prowling the halls of science. These days tigers are often confused with bugs (faults, flaws, or defects). Energy science has big bugs and small bugs, but all those bugs are as mice in comparison with the Big Tiger in the room: the attack on cause and effect.

In every scientific discipline, the greatest accomplishments have been the establishment of the causes of things. Science is a road with thousands of these discoveries, simple and complex, large and small. The discovery of what is causing the rusting of metals, or the source of interference to a radio signal, or why bridges are falling down, all are integral parts of the march of science through the ages. Cause and effect were conceived by the ancient Greeks, thrived under Sir Isaac Newton, and suffered under Einstein. The process can be stated as a simple cycle:

$$\text{Cause} \rightarrow \text{Distance} \rightarrow \text{Effect}$$

In other words, things have causes. Effects have causes. Cause and effect are not the same; a cycle of cause and effect must have

[111] R. Feynman, *Character of Physical Law*, Penguin Books, New York, 1992, p.70.

[112] G. E. P. Box, Science and statistics, *Journal of the American Statistical Association*, **71**, 356, 1976, 791–799.

distance to separate the cause and the effect. That allows you, when you want to discover why something happened, to trace it back to its cause. It might not always be easy, but it is available.

Modern quantum mechanics say that this is not true, and that something can be both the cause and effect of the same thing. They call this superposition, the same word they invented to describe a physical object "that can be in two places at the same time." Their denial of the existence of valid sequences is a serious illogic and, if not rectified, will keep the door closed to the discovery of new energy sources.

Graffiti keeps appearing on the walls of subway trains. Someone caused it. Lights keep burning out on the second car of those same trains. Something caused it. It has always fallen to the scientist to discover causes in order to provide better solutions to obstacles arising from life, work, or the environment. Science responding to these challenges has been called applied or experimental science. This includes the agronomist who improves the quality of soil and crops, the astronomer who plots the course of a meteor, or the computer engineer finding a better way to connect billions of people to the Internet. They have always been governed by the basic laws of classical physics, i.e., **things have causes, and exist in sequences.** If A caused B, then A occurred before B.

In the 17th century, there occurred a Scientific Revolution, led by Isaac Newton, British philosopher and scientist, born Christmas Day 1642. Newton created the subject of calculus and then used it as a tool to build the laws of motion. For the first time, he demonstrated conclusively that the movement of the sun and planets followed the rules of physics. He discovered gravity. His book "The Opticks" opened doorways of understanding into the properties of light. All these remarkable achievements had at their core the concept of cause and effect, and Newton introduced these concepts not only to science but to the society at large. For the first time, common people could appreciate the dynamic of cause and effect. It was a big piece of freedom Sir Isaac Newton brought to planet Earth.

We need to apply the principle of **Cause → Distance → Effect** to the task of finding new electricity sources. A scientist could interject himself/herself at any point on this cycle.

Lightning is an easy place to start since it is nature's most available showcase for huge electrical currents. Scientists could investigate basic causes of electrical potential that give rise to those huge voltages. Or they could investigate what is going on in the space between cause and effect during a lightning event. As will be seen in this book's final chapter, effective lightning research has been largely blocked by the same quantum mechanics confusions discussed earlier.

Cosmic rays may be another fruitful source of energy data. For 80 years, we have known of the existence of these high-energy rays. We have known they originate far away in outer space, though the exact points are obscure. Most importantly, we know they consist of very small wavelengths and that they pose a radiation risk to astronauts and electronic equipment that leave the Earth's atmosphere, yet those on the Earth's surface are apparently protected from them by some combination of the Earth's atmosphere, gravity, and magnetic fields. Why is that? And what is going on between the source of these rays and us? How do these "rays" make it across such vast expanses of interstellar space with enough oomph to harm an astronaut?

When first discovered, cosmic rays were correctly identified as a range of electromagnetic radiation of very high frequency and tiny wavelengths, but this is no longer the case. Quantum mechanics today refuse to even refer to them as rays. (A cosmic ray that is not a ray makes about as much sense as chicken soup without the chicken.) So how did these cosmic rays evolve from being rays to being non-rays? Well, when they were rays, they were considered to be composed of small particles of electromagnetic radiation and these particles had mass. Now the quantum mechanics say they are composed of small particles of electromagnetic radiation that have no mass; and if they have no mass, they cannot be rays. Sound crazy? Yes, but this is one of the more important illogics to be found in modern science: fictional particles that "exist" but have no mass, insisted upon to prove someone's equally illogical opinion about energy and space.

If science insists that particles have no mass, there would be no footprint to trace back to their origin. So, there would be no point in looking. So, there is no surprise that, as mentioned earlier, no one has yet found any points of origin of these cosmic rays.

True science establishes the causes of things. We have reached a pretty point in science when imaginary theories stand in the way of observation and discovery.

Number 1 Energy Miracle Obstacle: *Energy Can Never Be Created*

> *"According to Quantum Mechanics, this law is true in all cases 100% of the time, except when it conflicts with a quantum mechanical theory."*

We mentioned earlier that as soon as you task a new scientist with creating a 21st-century Energy Miracle, the first thing he is assaulted with is: "Energy can never be created." That may be true, and it may not be true, but it is a heck of a depressing starting point for someone trying to discover a 21st-century Energy Miracle. It is like trying to teach a student how to ride a bicycle by having him recite over and over:

> Don't ride at night, don't ride in the morning.
> If riding at noon take heed this fair warning:
> Consider just walking or crawling instead,
> Lest you fall off your cycle and land on your head.

Let us look at the history of the conservation idea. In 1773 the French chemist Antoine Lavoisier, Founder of Modern Chemistry, was studying the combustion of different materials and noticed that when he changed the state of matter in his experiments by subjecting it to a chemical reaction, the quantity of the matter before the change (based on its weight) was the same as the quantity of matter after the change of state. From this he concluded that in a chemical reaction, **matter** is neither created nor destroyed. This became known as the Law of Conservation of Mass. About 70 years later in 1842, Julius Robert Myer proposed the theory, now called the Law of Conservation of Energy, that **energy** can neither be created nor destroyed. He determined this by using vibration to heat water. What that actually showed was that kinetic energy was capable of creating heat. As interesting as that was, it falls far short of proof for an all-encompassing absolute law governing energy. (When I checked last week, no one had yet measured the quantity of energy existing before and after a nuclear explosion.)

It was Einstein, another 60 years later, who in 1907 announced his discovery of special relativity and merged the preceding two laws together as the Law of Conservation of Mass–Energy. This posited that the total amount of mass and energy in the universe is constant. Unfortunately for him when researchers were measuring the amount of energy emitted during radioactive decay, the energy amounts varied widely for the exact same decay process where the amounts should have been the same. Neils Bohr challenged Einstein's theory as a violation of the Law of Conservation of Energy, fomenting an uproar that lasted until another quantum mechanic, Wolfgang Pauli, dreamed up a new particle to explain the difference. He named it a neutrino, "an invisible subatomic particle with no mass, which travels faster than the speed of light and interacts in wondrous ways during the process of radioactive decay." This unseen but ubiquitous neutrino was the quantum mechanical answer to the inequality of mass–energy that had been actually measured during radioactive decay and other nuclear reactions. Another conundrum this presents concerns the more recent quantum mechanical "discovery" that the universe is expanding. And how does it do that if no additional matter and energy is being created to expand it with? Quantum physicists themselves admit that basic quantum processes violate the Law of Conservation of Energy.[113] Perhaps most importantly, how much trust do we want to place in energy theories that are not based on an accurate definition of energy itself?

What does it mean to create energy? People who lived on Earth several thousand years ago would have been astounded to see bronze being made in the explosive blast furnaces required to mix tin and copper. They would have assumed energy was being created. People living in China in the 13th century witnessing the firing of the first guns with exploding black powder would have been certain energy was being created when they saw a projectile heading toward them at speed. And people living in the 20th century would have been utterly certain energy was being created if they had witnessed an atomic bomb explosion. All those people would have been correct.

As we have seen elsewhere in this book, energy is simply the change in position of particles in space. In the search for Energy Miracles, all we are trying to do is organize certain facets of the physical universe into patterns and mechanisms that will incite flows of particles from one location to another. Whether those **particles** are created or destroyed in the process is another matter and irrelevant to that task.

[113]Y. Aharonov, S. Popescu, and D. Rohrlich, On conservation laws in quantum mechanics, *Proceedings of the National Academy of Sciences*, **118**, 1, 2021.

What is relevant to the energy researcher is that:

1. Energy can of course, definitely, and unreservedly be created; and
2. New forms of energy creation are available and can be found or invented.

Observation and Energy Research

> *"I believe in evidence. I believe in observation, measurement, and reasoning, confirmed by independent observers. I'll believe anything, no matter how wild and ridiculous, if there is evidence for it. The wilder and more ridiculous something is, however, the firmer and more solid the evidence will have to be."*
>
> —Isaac Asimov, The Roving Mind

You want to know how to research energy? You want to start your quest to find an Energy Miracle? You do it by communicating with energy. And that does not mean having secret conversations with your rechargeable batteries.

When we think of "communication" these days, we often think of it in symbolical terms: instant messaging, Google, opinions, tweets, formulas. They are not the communication lines that are required for energy research. Research communication lines have to do with perception, and the essence of perception is observation.

You originate some questions or communications toward energy and observe and receive back some return communications or information from that energy. You have got to look and see. Communication is, in essence, observation. You have interest in energy, you communicate with energy, and although it may not talk back to you, you get some perceptions or data back from that energy. You have to be open enough to confront whatever comes back. Any way you want to rig it, it is observation that you need.

It has got to be a two-way proposition. That means you have got to really look and be willing to observe what is there. A one-way proposition occurs when you look at something with a fixed idea and are unwilling to receive any incoming perceptions back from the thing you are studying that may conflict with that fixed idea. The failure to do that is the reason so many of the theories of quantum mechanics drift off into dream castles that have no relationship to the physical universe. And yet it is only in the physical universe that we will solve global warming and find 21st-century energy solutions.

The Father of Lightning Research was a Swiss gentleman named Karl Berger. He learned more about lightning than anyone either before or after him. How did he learn so much? He persuaded some people to give him access to a couple of towers on mountaintops in the Swiss Alps and on any given day during the lightning season for 30 years, you could be sure to find him at one or the other of those towers, observing, studying, and measuring actual lightning. His results, though more than 70 years old, have not been supplanted.

You want to research about energy, you start off by observing it. The essence is communication and observation.

Of course, observation and experimentation require some work. It is far easier to go online and Google lightning than go out on a mountain and observe some. Trust me, there will be people who will attempt to discourage you from actually going out and observing things. But is not that the trouble with all of quantum mechanics? They do not want you to observe.

The big challenge is whether you dare to observe. And if you do, a great guide to getting started on your quest for Energy Miracles is the book *Principles of Electricity Containing Divers New Theorems and Experiments* written by Charles Viscount Mahon, FRS (The Earl of Stanhope) in 1779, predating quantum mechanics by over a century. His directions and diagrams allow you to duplicate over 60 of his basic electricity experiments.

Stanhope's book can be downloaded from the www.energymiracles.net website, along with other data that may help you on your quest.

Energy Two × Two

"He called for the magnets
They each came in twosies twosies.
Turbines and solar
They generate by twosies twosies
Nuclear and wind, even they toosies, toosies
Energy comes from two."

—The Modern Arky Arky Song

We are looking for new 21st-century energy sources. Energy consists of postulated particles in space, and we find whenever energy is generated, there will be two "terminals" and a line between them.

You can see this clearly in a DC battery where energy flows from one "**terminal**" to the other "**terminal**," so we will maintain that terminology. If one terminal were to disappear, there would be no flow of energy.

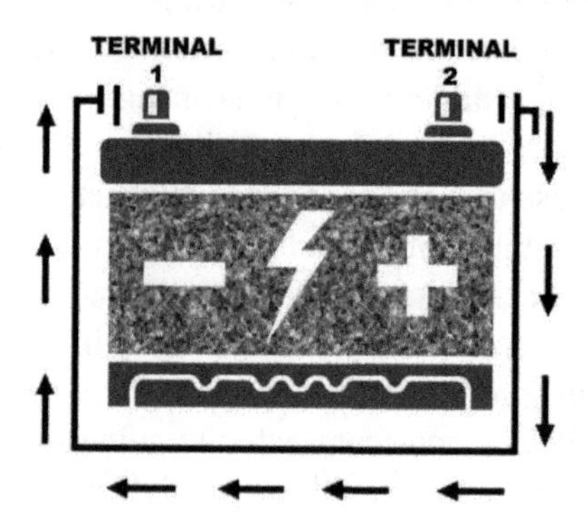

Figure 107 Two terminals are needed in a battery for energy to flow.

Two terminals must be established for energy to flow between them. Take lightning. There must be a storm cloud overhead with a massive charge build up (Terminal 1) and a location on the ground with zero electrical charge (Terminal 2). Minus either one of those, there would be no lightning flow of energy.

This mechanism is fully in play in the creation of electrical energy and magnetic force. It is also true for gravity and every other type of energy that has ever been observed. For energy to flow, there must be a terminal at each end providing the conditions that create that energy flow.

Figure 108 Two terminals needed for any energy to flow.

So, which do you think is of senior importance? The terminals or the flow of energy? If you said the terminals, you would be right.

Why? Because you need the terminals in order for the line of energy to be established; without the terminals, there could be no energy. (Although it may sound esoteric, it seems to be a fact that people who are unaware of terminals are reduced to seeing only lines. And those who are unaware of even chaotic lines are likely to become fixated on menacing or nonexistent particles. Welcome to the world of quantum mechanics.) Anyone can increase their awareness of terminals if they work at it. One way this can be done is simply by observing things.

When scientists were looking at the cause points and effect points of electricity, they made great progress in energy generation. When science was looking at the actual elements and processes, messing with them, measuring them, discovering their individual characteristics, and how they co-acted with other elements and processes, they made extraordinary breakthroughs. Now our most "advanced" scientists specialize in bombarding the dickens out of a few of these elements in particle accelerators "to see what happens." It is like trying to learn about girls by beating them and then studying the pitch and duration of their screams and the type of blood they shed. Whatever they "learn" from this, it is not going to give them any greater understanding of girls.

New sources of energy will not come from studying a confusion of little doodle-daddles running up against some thingamabobs. Twenty-first-century alternative energy sources will consist of new (innovative) pairs of terminals capable of generating a flow of particles or impulses between them.

Power of Simplicity in Energy Miracles

> *"Nature is pleased with simplicity. And nature is no dummy. Truth is ever to be found in the simplicity, and not in the multiplicity and confusion of things."*
>
> —Sir Isaac Newton

Science often makes the mistake of believing that whenever it is confronted with a great deal of complexity in some difficult situation, the solution has to be more complex. Yet scientific discoveries of great value almost all involve simplicity. The greatest contributions of Galileo, Newton, and Maxwell all took vast complex situations shrouded in mystery and unknowns and reduced them to understandable simplicities.

"Simplicity" here does not mean rudimentary or primitive as can be seen in the following example. The Internet may qualify as being the greatest creation of our generation. It took dozens of completely different forms of communication, from phones, to beepers, to faxes, to letters, to telexes to telegrams to emails, to radio and television and put them all onto a single platform. Where before great complexity was involved in the simple action of placing a phone call from one country to another, the Internet became a seamless carrier of global communication. Where previously, language barriers created thousands of obstacles to communication, the Internet made it almost easy for native English and Chinese speakers to communicate with each other across a distance of thousands of miles. We would not even talk about Internet commerce and shopping or the management tools it enables. The Internet is not a "simple" operation; hundreds of thousands if not millions of personnel and thousands of complex technologies are involved. Simplicity here means a single invention incorporating a myriad of discordant processes into one easily accessible and usable system.

Any fool can come along and say, "Well, let's see, you've got things nice and simple. Now let's make it more complicated. In order to study new sources of energy production you must first master the Siamese alphabet backward. And once you've got that you've got to learn $E=MC^2$ divided by the square root of river rats. And if you can't understand that, then you have no business studying this subject and we don't want anything to do with you."

Modern science did not originate this tendency. The quantum mechanics have just specialized in it. They get more complex and more complex and never dream for a minute that they would better look for simplicities. But simplicity is where the Energy Miracles are to be found.

A Scientific Datum Should Embrace Other Data

Anyone saying a wormhole is as important as Newton's first law of motion has missed the essential point of data.[114]

---

[114]The speculative thing inside the cube in Fig. 116 is a wormhole.

All data did not hatch out of the same pig. There is data, and there is DATA. The essential point of scientific data is that **some data embrace other data**. The opposite, a monotone value, means that every datum has the same importance or value as every other datum. We have mentioned a few of the great discoveries of Isaac Newton in this book. They are great because they embrace so many other data. His first law of motion[115] is valuable because it embraces thousands and thousands of other data related to moving books, driving cars, digging ditches, walking to the store, and flying planes.

Louis Pasteur, the French biologist who is famous for his breakthroughs in the cause and prevention of disease, promulgated a simple datum that embraced countless other data: Germs cause disease. That one datum solved thousands of non-optimum situations, resulted in improved hygiene, pasteurization, water treatment, preservation of food, vaccination, saved millions of lives, and was responsible for improved living conditions for hundreds of millions of others. That is an example of a simple but well-grounded discovery embracing a huge number of other data.

A scientific law, datum or equation is as valuable as it embraces other data. A Quantum Wormhole is valueless because it embraces no other data.

The basics of energy given in this book embrace a myriad of different types of energy creation. Studying, observing, and gaining a strong understanding of these basics will put you firmly on the road to the discovery of Energy Miracles.

The Misunderstood Word and Energy Miracles

With the disappearance of real dictionaries in the Internet age, a scientist's ability to smoothly skate through scientific texts is greatly hindered. By "real dictionary" is meant one that gives every sense of the word in its definitions, including the archaic and any specialized uses, together with examples and sentences and etymology.

Any word or symbol a student or scientist contacts that is not fully understood stands in the way of his or her being able to fully grasp the material he or she is trying to learn.

---

[115]That an object will remain at rest or in uniform motion in a straight line unless compelled to change its state by the action of an external force.

Figure 109 A real dictionary.

Take the sentence: "Should we be investigating the nudiustertian power outage?"

If you have only a foggy concept of the request being made, it will be because you do not have a complete definition for one of the words. It might become considerably clearer once you discover the meaning of nudiustertian.[116]

Lack of misunderstood words also establishes aptitude. If you find someone who has just graduated from a cooking class but cannot boil an egg, you will find many words he either did not understand or wrongly understood. Misunderstood words can make a student feel blank or washed out. (In that case, he or she should go back and clear the word that was misunderstood just before he or she started to blank out on the material.)

The people involved in the resurgence of innovation into new energy sources will need a clear understanding of the basic terms involved. It is a simple process, but clearing up the definitions of all the words in one's subject is vital for scientists who want to be whiz kids and able to fully apply the materials they are studying.[117]

Anyone on the quest for an Energy Miracle would do well to ensure they had a full conceptual understanding of the following terms, all of which have been defined somewhere in this book:

---

[116]nudiustertian—adjective. Referring to the day before yesterday.

[117]The misunderstood word is the third and most important barrier to study discovered by Ron Hubbard. His *Barriers to Study Booklet* can be downloaded free from the Internet.

energy
static
kinetic
matter
space
time
poles
dichotomy
magnetism
mathematics
empirical
generate
generator
line
terminal
flow
dispersal
ridge
explosion
implosion
electricity
engineer
pressure
universe
vacuum
knowledge

An updated list of terms together with their definitions is available on the website www.energymiracles.net.

The Art of Demonstration

If you wish to formulate a useful bit of law,
If you care to save yourself from falling down a maw,
If you crave avoidance from seeking pins in straw,
Better figure out a way to demo or to draw.
Demonstrate; demonstrate...the saving grace of science.
Demonstrate; demonstrate, for student full-compliance.
Demonstrate; demonstrate means SHOW HOW YOU CAN USE IT.
If you cannot demonstrate, you might as well just lose it.

There is a basic concept in science that if you are not able to demonstrate something, you do not understand it clearly enough to make it work. To demo something could mean to make a detailed scale model of it, but the process can be short-cut to roughly creating it in clay, drawing it, or even using bits of paper clips, coins, rubber bands, and pencils to satisfy yourself, your professor, or your boss that you can actually apply the data.

Scientists these days erroneously believe that a mathematical formula is visualizing. It is not. Mathematical formulas are merely symbols.

There is a breathtaking movie by the late filmmaker, Felix Greene, who went to China in the early 1960s to record the challenges facing a proud but hard-pressed people working to build a country. His favorite clip shows a team of Chinese engineers charged with constructing the foundation for a causeway that was meant to span a 5 km strait. To make a causeway, you need rocks, millions of them, but with no mechanical equipment to assist them, they started the project by dropping one rock at a time from tiny sailing skiffs.

Figure 110 Frame 1: The original system dropped one rock at a time. Frames 2 and 3: Engineers demonstrate their proposed solution. Frames 4 and 5: Plan in action—successfully dropping 1000 rocks at a time within 4 seconds. Frame 6: The completed causeway. Images included with permission of Greene family.

They worked that way for months, but as the strait closed, and the width of the channel narrowed, the currents running through the strait increased until the boats were being shot through at 40 knots, making it impossible for the workers to drop their rocks in the 4-second window the boats were in position. They needed to innovate.

A lot of thought and energy went into figuring out a solution and then demonstrating it: first on paper, and then in the scale model construction. And largely because those demos were detailed and accurate, potential pitfalls were avoided, and it actually worked in real life. Those engineers knew that if they could not demo it out first on paper, and then with a scale model, they would not be likely to get it right. The reason? Demos are very likely to show up incomplete planning, wrong assumptions, confusions, or omitted parts. So, if you cannot demo out or draw a concept or idea to show how you could actually do it or use it, you need to study the item or problem more.

The film pans from images of exhausted workers dropping one rock at a time, to the successful solution where 1000 rocks were being dropped at one time. See Fig. 110.

The great William Thomson (Lord Kelvin) once said, "I am never content until I have constructed a mechanical model of the subject I am studying. If I succeed in making one, I understand, otherwise I do not."

Science and Understanding

Quantum mechanics tell you the closer you look at something and communicate with it, the less you will understand it. This is one of the larger falsehoods they have introduced to the world of science. It is difficult to find statements that are further from the truth. Understanding in science (or anyplace else, for that matter) is achieved mainly through observation and communication. Without communication, no understanding is possible. Any scientist worth his white lab coat has unflaggingly followed the basic law that the more he communicates with something, the more potential there is to understand it. Without communication, there is no understanding.

With communication, one may achieve an appreciation for the intricacies of a thing and for its reality. The more that an observer duplicates an object's composition, structure, and reactions, the greater will be his or her understanding of that object. It is also evident that with communication comes affinity. Someone seeking to really understand an object or a subject or even another person will inevitably develop some sympathy for that thing. It is not just a coincidence that almost all of the great electrical inventers on the top 42 list were not professionals, nor were they primarily motivated by money. Their studies were spurred on by an affinity and curiosity for their subject. For energy. Their work was often conducted with enthusiasm in their spare time over protracted periods (including all night long). The basic aspect of affinity is the ability and desire to share the space of that which is being observed. The ability and enthusiasm for doing this with something increase one's understanding of the thing—vital in science. There is no absolute level or degree to any of the preceding factors. They exist as gradient scales. More and more; greater and greater; or less and less.

What is required to achieve a given level of understanding is to achieve a commensurate level of communication and observation. Do not expect a quantum mechanic who has brainwashed himself into believing that communication with an object ruins his chance of understanding it, will ever come up with any useful source of alternative energy.

No communication equals no understanding.
Any understanding requires some communication
More understanding requires more communication.
The most understanding requires the most communication.

Figure 111 How science achieves understanding.

Getting Started on the Path to the Energy Miracles: A Study Guide

"The student who uses home-made apparatus, which is always going wrong, often learns more than one who has the use of carefully adjusted instruments, to which he is apt to trust and which he dares not take to pieces."

—James Clerk Maxwell[118]

[118] J. C. Maxwell, *Maxwell on Molecules and Gases*, edited by E. Garber, S. G. Brush, C. W. Everitt, MIT Press, 1986, p.113.

This book has dealt with the subjects of climate and power generation and their inevitable interconnections and pointed out the need to come up with a 21st-century system of power generation—an Energy Miracle. It even provides an international mechanism to jump-start the process. We have tracked the cessation of scientific advances in energy production to the adoption of the theories of quantum mechanics. We do not here make any claim that those theories are correct or incorrect. We only point out that when they were adopted, innovations in energy production ceased. Conversely, we have identified specific scientific principles that were nullified and abandoned at the time electrical innovation ground to a halt and suggested that taking a fresh look at those ideas was a logical way to throw the door open to 21st-century Energy Miracles.

In the following pages will be found a roundup of the basic definitions and actions of energy. Once he or she is familiar with these, the seeker of Energy Miracles should begin a study of the scientists listed in the "Top 42 Electrical Discoveries" in Table 2. Read about them in books like Percy Dunsheath's *History of Electrical Engineering*, Faber and Faber, London, 1962, or *Fleet Fire: Thomas Edison and the Pioneers of the Electric Revolution* by L. J. Davis, Arcade Publishing, 2003. Get into their heads and follow their trains of thought as they studied the relationships of electricity and magnetism to other forms of energy. Most of them, once the spark was in them, threw themselves with a fury into their electrical researches. Some did their researches after work in the evenings and on weekends. Others devoted a lifetime.

Then study their lab experiments and seek to duplicate them. Most of the greatest advances were achieved with inexpensive and easily constructed equipment. Cork, iron balls, glass bottles, silk thread, and a kite were all the accessories needed by Benjamin Franklin to perform the electrical experiment that resulted in one of the most important discoveries in energy science. The Earl of Stanhope's 62 electrical experiments can be duplicated by anyone and are eye openers. The experiments of Michael Faraday and James Clerk Maxwell are vital. These are all available for download at www.energymiracles.net.

When you have done that, study Tesla's work.

The Basics of Energy, Space, and Time

You want a head-start in the Energy Miracle Challenge? You have got it. Study and understand the basic definitions and descriptions of energy given in this book. A short roundup is given here, including definitions and principles from the works of Ron Hubbard. Known principally as a writer and philosopher, Hubbard was also a sea captain, aviator, horticulturist, photographer, musician, and humanitarian. Perhaps his most underappreciated contribution was to the science of physics. He attended one of America's first classes on nuclear physics while studying at George Washington University in 1932. The subject was then called Atomic and Molecular Phenomena, and the class sparked an interest in the mechanics of energy that he maintained his entire life. I was privileged to know him, and his interest in science and the subject of energy inspired my own.

As far as energy is concerned, Hubbard saw clearly that the highest order of action in creating electrical flows was to locate or establish in space, things which would discharge from one to the other. You do only that, and you have got electricity flowing.

We are living in a time when authoritarianism and dogma obscure important parts of science so thoroughly that despite the paramount need for new, clean, abundant, and cheap sources of energy, it requires extraordinary resolution to research in these areas. Achieving a successful result in the quest for an Energy Miracle requires a true scientific methodology, and Hubbard was an avid proponent of such approach. Up to now the individual researcher in search of Energy Miracles has been buried under the weight of abstract ideologies, colliding at every hand with over-financed research interests.

Unsurprisingly, despite billion-dollar particle accelerators and colliders, there have been no advances in new types of energy production.

Statics and kinetics

Physics can be said to be a study of statics and kinetics, but what does that mean? Modern physicists will look at a small cube sitting on a lab table and say, "It is not moving, so it is a static." Not so fast.

That thing has a half dozen different motions on it already if it is on the surface of the Earth. There is the motion of the Earth moving around the sun, which is a variable motion of 30 km/s (67,000 miles per hour). Then there is the spinning of the Earth on its axis, which it does at the rate of 460 m/s (1000 miles per hour). There is also the 200 km/s (448,000 miles per hour) that the Earth is moving around the center of the solar system. And if you want to look inside that little cube, you will find the molecules and atoms of which it is composed moving around like crazy. You would have to be a better magician than a physicist to call something with that much motion a static, for a static, by definition, would be something that had no motion in it. More specifically, as covered earlier, a static would be something that had no mass, no motion, no wavelength, and no location in space or in time.

A kinetic, on the other hand, is a flow of energy particles, which can exist at any wavelength from one over infinity down to the wavelength of infinity. But the moment you say it has zero wavelength, it becomes a static and no longer has a wave motion.

The successful seeker of Energy Miracles will need to study the difference between statics and kinetics.

Energy

Energy consists of postulated particles in space. We do not quantify them. They can be big or small, but they are particles, they have mass, they require space, and they are either in motion or not in motion. Energy is action, and it can also become an object. There are three varieties of that happening. There is a **flow** and there is a **dispersal**, and there is a **ridge**. These are the three kinds of energy, the three actions of energy.

Energy's three characteristics are shown in Fig. 112.

A **flow** (marked A in Fig. 112) is a directed stream of particles. This could be a smooth wave, a sine wave, or a much noisier or complex wave. It could be of any wavelength or amplitude. All flows can be graphed. They include a stream of water, a beam of light, a rarefaction/condensation wave (which passes through air), or a particle (such as X-ray) emitted from a machine. A straight line is inevitably a type of wave, because a wave is essentially a path of flow.

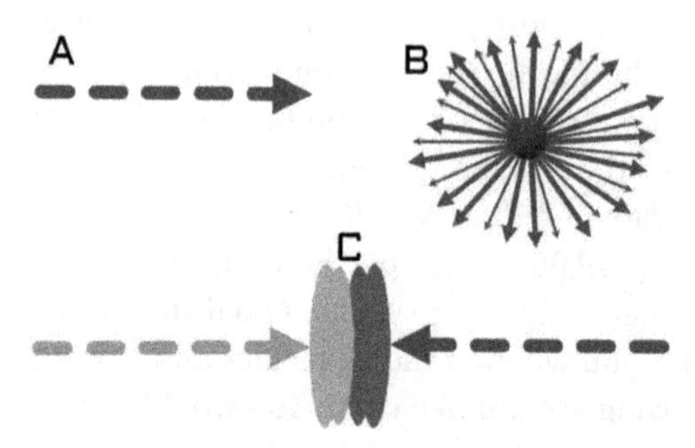

Figure 112 The three basic energy characteristics: (A) flow; (B) dispersal (explosion); (C) opposing energy flows creating a Ridge.

A **Ridge** (marked C in Fig. 112) is the second type of energy. Standing waves are an example of a ridge. Two cars running into each other at speed form a "standing wave" at the immediate point of impact. Two flows hitting up against each other form a continuing state of matter. If you were to take two buckets of water and throw them at each other, you would have two flows meeting in mid-air, and if you had a stroboscopic camera to capture that instant, you would be able to study the pattern they made. The two flows hitting each other will make a wall.

Mountains are an example. Mountains are created when gigantic flows or dispersals pushed against landmasses bringing new ranges of high peaks into being that lasted long after the original flow that created them. Momentary ridges can be seen in the atmospheric or electrical interferences to radio or network signals that result in static or hums or just poor reception. A hydraulics engineer will tell you that it is not just the makeup of the medium that creates the waves. Waves form at any place where two fluids or substances of different densities meet. Ocean waves are formed from the denser water impacting the relatively less dense air. When a car runs into a cat, the cat splatter on the road can be considered the result of a ridge—though a not very aesthetic one.

The third characteristic of energy is the **Dispersal** (marked B in Fig. 112), the (often explosive) distribution of matter or energy outward, spherically, from a center point. It is a specialized kind of flow—a multiple flow. Examples are a nuclear explosion or the day-to-day action of our sun. Under dispersals, we also classify implosions. An implosion could be called a reverse dispersal, where all the surrounding matter is sucked into a single point. These are actual descriptions of the behavior of particles in space. A black hole is an example of that as is an out-of-control submarine being crushed by the great pressures existing far below the surface of the sea.

Ridges are best formed when two flows hit each other. But the impact of two dispersals hitting each other can also make a ridge, as will the combination of an implosion and an explosion in such a way as to achieve a constant turbulence of electronic flow. All these and more create ridges.

A short examination of these three types of energy will show a multitude of interactions: A flow of heat against a piece of matter (such as a log in a fireplace) will cause a dispersal when that piece of matter absorbs enough heat to "burst into flame." Two armies exploding out at each other (two dispersals from different sources) can create a ridge at their point of contact when men and equipment are built up on each side of the contact point with neither side being strong enough to overrun the other. The Berlin Wall was such an example.

Matter

Matter and solid objects are grouped particles. Matter is generally thought to exist in one of three states, depending on the degree of particle condensation: solids, liquids, gases.[119]

Figure 113 Different states of matter depend on density of particles.

To give themselves a backdoor out of having to explain failed theories, quantum mechanics now say that over 99% of the matter in the universe is in the plasma state, which is not even a state of matter. (See footnote 119.) For the past 20 years, quantum

[119]Textbooks too heavily influenced by quantum mechanics will add plasma to that list, but a plasma is not a different state of matter. It is merely a gas with an electrical charge imposed upon its particles.

mechanics have also been saying that the universe is composed of 68% dark energy and 27% dark matter. If true, that means Einstein's relativity theories applied to less than 5% of the actual universe. Notwithstanding the above, where Einstein got himself into the most conflict with reality was when addressing the other two components of the physical universe: space and time.

Space

Simply stated, space is a viewpoint of dimension. Whatever dimensions you are viewing, that is your space. You say: "I am here looking in a direction." To have two-dimensional space (length and breadth but no depth), you need at least three points.

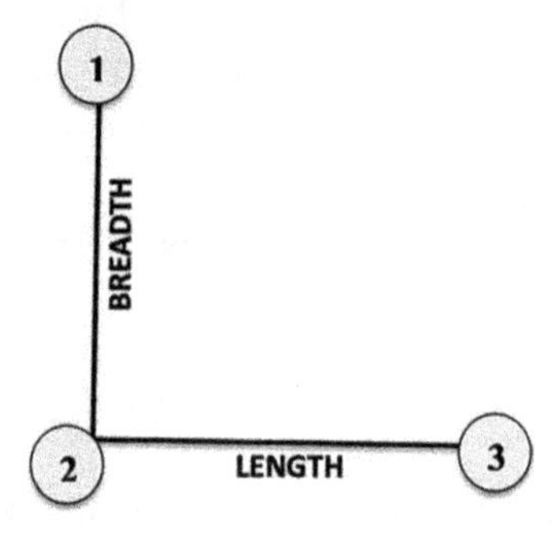

Figure 114 Three points are needed for two-dimensional space.

The eight corners of a room are an example of three-dimensional space.

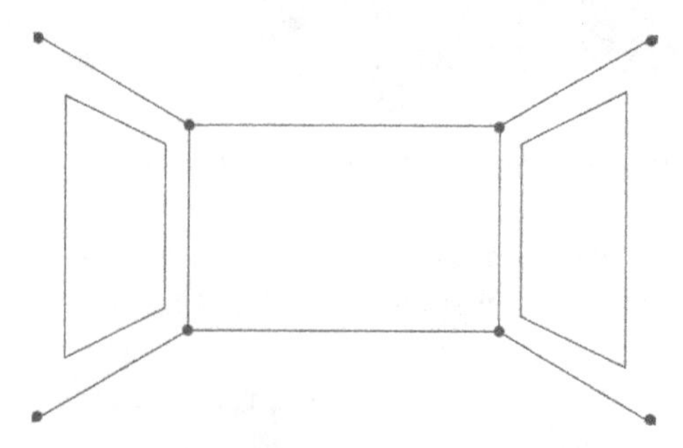

Figure 115 An example of three-dimensional space delineated by the corners of a room. The eight-cornered room still only possesses three dimensions: length, width, and height.

If you put only two points out, you would have one-dimensional linear space.

Different observers view different dimensions. They are all equally valid. Mathematicians have taught us about one-, two-, and three-dimensional spaces. Those observations are accurate, and they cover all the spaces in this universe. Although Photoshop and any of the new science drawing apps are able to produce some pretty nifty drawings, there is not one of them that cannot and does not exist within three-dimensional space. Figure 116 shows an imaginary quantum wormhole in delusional n-dimensional space. Awesome graphic, but as can be seen, it fits very handily into the three-dimensional space of the cube.

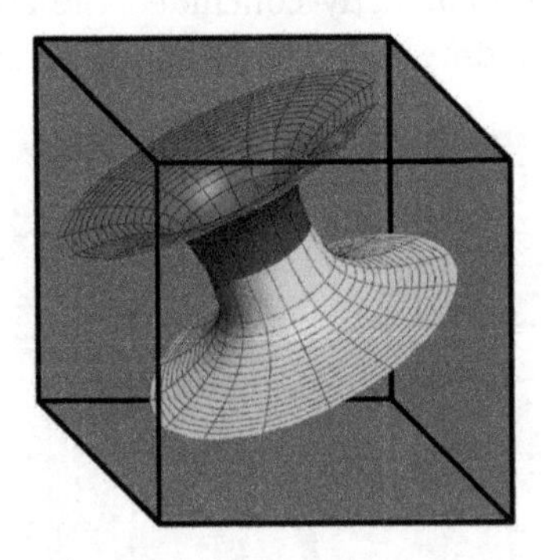

Figure 116 A quantum wormhole in n-dimensional curved space fits very well into the ordinary three dimensions of the cube.

Space and Dimension
(no dimension = no space)

Dimension	0	1	2	3	
Minimum # of points or particles	1	2	3	4	
Space is a viewpoint of dimension	•		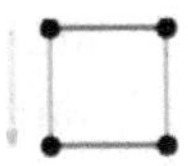	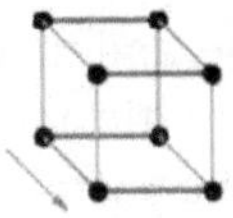	X Y Z
Space?	No	Yes	Yes	Yes	

Figure 117 With a single point there is no dimension and no space.

The number of dimensions and the number of points demarking them are critical to many areas of science. As shown in Fig. 117, a single point has no dimension and hence no space. As shown elsewhere in this book, a single point also has no charge. This gets the quantum mechanics stuck when trying to theorize the properties of their imaginary dimensionless single particles.

Time

Einstein was caught up in the conundrum of whether time was "relative" or not. He pondered whether there was a big atomic clock up there in the sky ticking away at an unvarying tempo by which everyone danced. He correctly concluded the answer to that was negative, and that all time was relative to the person observing it.

Figure 118 There are many systems of time. In this depiction of quantum time, the circles represent an infinite number of interconnected clocks.

The preceding "quantum time" depiction shows all the clocks connected to each other, but in fact, there are a huge number of systems of time. You can say "the sixth birthday of my cat" or "3 years after he graduated" or "just before the accident" or "five generations ago" or "at the end of the uranium's half-life." These are all valid systems of time, and they have little or nothing to do with clocks.

Time is not a ticking clock, and clocks create neither space nor time. Time is the postulate that matter and energy will persist. Time marks the co-action of particles or the change in position of particles in space. The common denominator of time is change, and change is the primary manifestation of time. If particles are absent or are unmoving, you have no time. A particle moving, changing its position in space, such as the dog in Fig. 119, manifests time.

Figure 119 Time marks the change in position of particles in space.

Rarefaction and condensation

This is one you could very easily miss. The fifth key to the Energy Miracles is the means or mechanism by which energy propagates through a medium. This has been a forgotten element of science since energy particles were mis-defined and the concept of a medium abandoned. Earlier in the book, we provided references to earlier science texts, which unequivocally stated that the principle of the propagation of light waves was the same as that for water waves. The LIGO folks who won the Nobel Prize in Physics a few years ago for their observation of light waves confirm that light waves behave just like water waves. See footnote 96.

The way not to understand light waves is to tie one end of a rope to a tree and then flick the other end. That is not how electromagnetic waves propagate. It they did, you would need some kind of rope between the local TV transmitter and the TV antenna on your roof, and you would have to hire somebody at the station to keep flicking that rope.

Electromagnetic waves (including light waves) move through space utilizing the condensation/rarefaction mechanism. In this terminology, the ridges we introduced earlier can be understood as condensations of matter or energy while flows and dispersals can be seen as relative rarefactions. See Fig. 120.

The condensation/rarefaction mechanism is ubiquitous. It is vital to a full understanding of the double-slit experiment, as explained earlier, and is evident in almost every phenomenon described in Gerald Pollack's remarkable book *The Fourth Phase of Water*. What is termed an exclusion zone (EZ) in that book is a condensation, a compaction of water molecules. The surrounding water consists of more rarefied molecules allowing for more active flows and dispersals. Pollack found that it is the radiation entering a system that drives charge movements (energy). This holds true in both water and in thunder clouds and has even been proven to be the prime mover of Brownian motion. Pollack is an old school scientist, who politely decries the modern quantum approach of abandoning the pursuit of simplicity. He strives for simple unifying truths, and it is clear he has found some in a book that has been described as the most significant scientific discovery of this century.

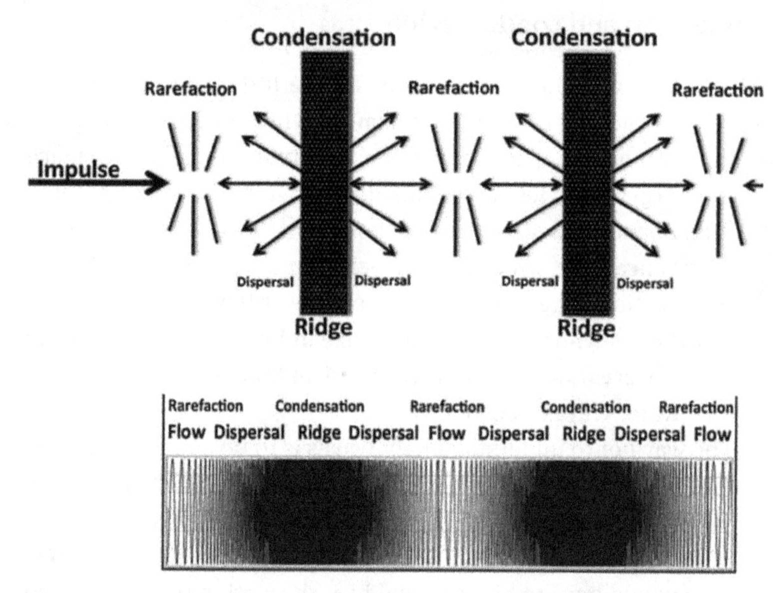

Figure 120 Energy progressing from left to right. Ridges are condensations; flows and dispersals are more rarefied particles.

The basic patterns of energy can be seen in Fig. 120. In a condensation/rarefaction wave going down a copper wire, what is happening is the electrons are rarefying and condensing. The electron does not move very far or flow down the wire like a drop of water. The electrons rarefy and condense. Figure 121 shows what this has to do with particles. In the top frame you can see particles, particles, particles all over the place, evenly distributed. These particles are just going along quietly; there is nothing happening to them yet.

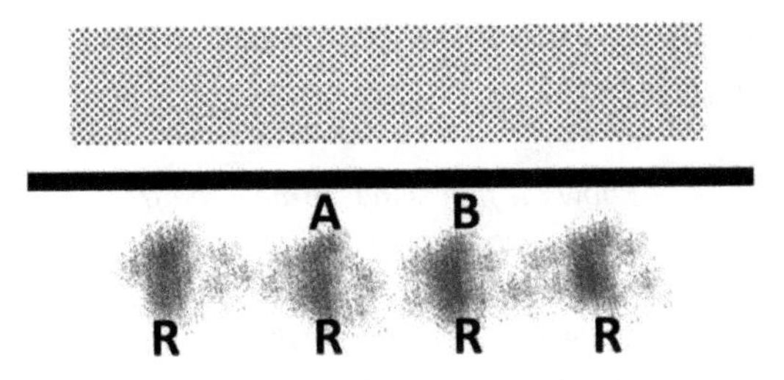

Figure 121 Top: Particles unaffected. Bottom: Particles in a wave.

Now in the bottom frame, we have put a wave through those particles. You can see them grouping. There are embryonic ridges at the parts marked "R." Those ridges are just condensations of particles. The rarefactions are the places where there are fewer particles. A complete wave is the distance between the nodes (marked A and B). Realize those are spheres you are looking at—three-dimensional spheres.

If you examined, stroboscopically, the particle flow of a rarefaction condensation flow, you would get minute patterns, which would demonstrate that there were, at any given instant, rarefactions and condensations taking place, and that some of the particles between the rarefactions and the condensations were expanding suddenly and some of the particles were crashing in, and the pattern of particle action would give you a pattern that you see approximated in the bottom frame of Fig. 121.

This also explains standing waves. Supposing we got this rarefaction condensation wave going good enough and heavy enough and then opposing flows just ground into it and stopped it. That pattern of the Rs in Fig. 121, if closely examined, would become the pattern, more or less, of the condensations in Fig. 120. Those ridges would stand.

The condensation/rarefaction mechanism describes not only the pattern of an electrical flow down a wire, but also electromagnetic flows through space. It is also the pattern of a galaxy and the pattern of an atom. It seems that almost everywhere you look in the universe you find evidence of this mechanism.[120]

---

[120]More information on condensation/rarefaction can be found in the next section on lightning and at www.energymiracles.net.

Final tip to those who wish to discover an Energy Miracle

"Know the first thing we did wrong?
Went along with the quantum song.
Keep your eyes on the prize. Hold on.
Know the first thing we did right?
When we challenged the speed of light.
Keep your eyes on the prize. Hold on."

—Traditional Classical

Figure 122 Persistence overcomes resistance.

Persistence on a given course is one of our greatest abilities, appreciated in all professions, not least in the search of an Energy Miracle.

The search cannot be quickied. That means it cannot be brushed off or done so fast that it ends up not getting done at all. I can clean a room rather quickly, picking up the dirty clothes, and making the bed in a mere 5 min. My wife, on the other hand, will take 2 hours to clean the same room. The results are not the same. Not even close. She gets into every nook and cranny with her dust rag, moves furniture to get behind it, gets down on her knees to scrub the floor, and uses a ladder to get at the high ledges. None of those actions were part of my procedure, and each time I witness how great the room looks, feels, and smells after her version of "cleaning," it brings home the embarrassing fact that she actually cleaned the room and I did not.

This is a vital lesson for those seeking an Energy Miracle. In the first place, you need to get a good grasp of the basics of energy. In the last section, there is a suggested course of study. Take care you do not quickie it. One of the suggested steps is to conduct the experiments given in one of the books you can download from the www.energymiracles.net website. This is a totally doable assignment, but it cannot be done in 5 min. Nevertheless, I can pretty much guarantee that somebody is going to skim through that book of experiments in 5 min and decide "I have done it." No, they have not.

Currently there is an app where you can look up a book and be provided with "the gist of it" in a 15-min read. And if you cannot bother to read even that, the girl imprisoned in the app will read it out to you. Imagine, four experts spend 5 years researching something important, process all that data, discard 95% of it, and further refine the remainder down to 300 pages, that being the least space that could encompass sufficient information to give the reader a reasonable idea of the subject. Then some guy comes along and says, "Hey, people do not have enough time to read a 300-page book. I can reduce it further down to a 15-min wink to give them the gist of it." The "reader" then reads the book so fast he ends up not having read it at all.

My interest in electricity began 65 years ago, when in the second grade I determined to better understand what force was coming out of those two-slotted plates in the wall and planned an experiment to investigate it. That night I borrowed a metal hair pin from my mother's cosmetic table, and the next day in school, just before lunch, I bent it to just the right angle and inserted the ends into the two slots. Bingo. In the ensuing explosion, I found out way more than I intended when the electrical system of the entire school went dead. My fingers burned for a month, and the smell of my blackened flesh stayed with me for years. I do not recommend you try something like that, but there are some experiences you have to undertake personally to grasp what energy is, and become a master at creating new sources of it. That includes understanding each of the words listed in the earlier sections of this chapter and getting your hands wet playing with actual electricity. It will definitely not take anywhere near 65 years to make you an expert, but neither can it be done in 5 min.

You are sure to run into barriers in your study, barriers in your experiments, and (finally) barriers in creating new sources of energy. But you will win if you persist. What does that mean? It means keep at it. Keep creating it. Keep moving on the path toward the Energy Miracles. That goal may be the most important goal on Earth at this particular time. You can reach it. It is there for the taking, if you only persist.

Keep your eyes on the prize and hold on.

Good luck.

Chapter 10

Paths to the Energy Miracles

The five keys to the Energy Miracles are found in Chapter 6. Almost every one of the greatest electrical discoveries in history (see Table 2 in Chapter 5) was based on those five keys.

Figure 123 Paths to the Energy Miracles.

Energy Miracles are within reach with an understanding of those five keys. Using them as a guide, there are many avenues where they can be profitably applied. Some of the most exciting ones are as follows:

Thermoelectrics

For seekers of an Energy Miracle, thermoelectric is a great sounding word: It means "electric energy from heat," and in fact, in 1821

Energy Miracles: The Global Warming Backup Plan
H. B. Glushakow

ISBN 978-981-4968-18-8 (Hardcover), 978-1-003-28442-0 (eBook)
www.jennystanford.com

the Estonian physicist Thomas Johann Seebeck discovered that a temperature difference between dissimilar metals would produce an electromotive force (voltage) around wires that would drive a current through a circuit.[121] He also found that the voltage varied with the degree of temperature differential, so that the greater the difference in temperatures, the greater the energy that could be produced. In a laboratory at room temperatures with ordinary materials, such voltages are measured in microvolts. But temperature differences cause great storms to form, winds to blow, lightning to flash, and huge hailstones to fall. It is clear that the difference in temperature between two objects has the potential to directly create electrical energy. The temperature differences create a kind of plus and minus—much as the difference in voltage that creates a lightning strike. That potential could be harnessed.

This relationship, called the Seebeck effect or thermoelectric effect, is the physical basis for a thermocouple, which is often used for temperature measurement. But aside from thermocouples, the concept has been largely ignored over the past 165 years.

Thermodynamics has given way to geothermal (heat from the Earth), which includes methods to convert heat into mechanical energy by bringing steam up from under the ground to power electric generators or using underground hot water to warm homes.

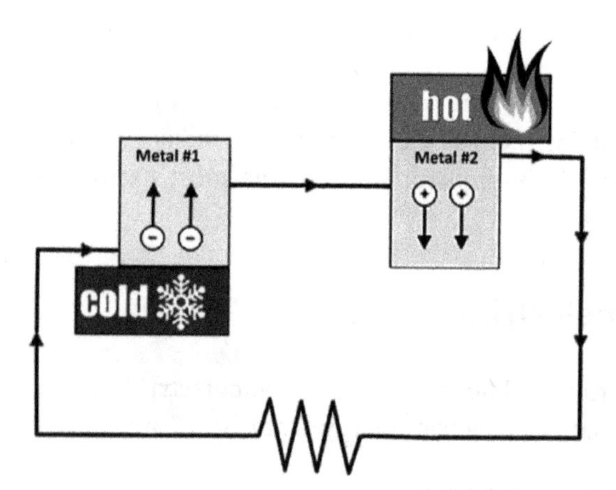

Figure 124 Thermoelectric energy.

[121]Alessandro Volta discovered the same phenomenon 20 years earlier, but it was nevertheless named after Seebeck.

What if we had a more or less unlimited source of extreme heat that, when juxtaposed with another object of much cooler temperature, could directly create an electrical current? And what if we could do it without the need for digging up coal, burning coal, using vast amounts of water for steam to run a turbine, and yet produce that same electrical current?

That heat source exists. Scientists estimate that the heat at the planet's core could be as high as 6000°C. That is as hot as the surface of the sun. It is 6371 km (4000 miles) to the center of the Earth, but you do not have to dig anywhere near that far to encounter some really serious heat.

Figure 125 is a diagram of the cross-sectional area of the Earth. The Earth's crust is 35 km (21 miles) thick on average. At the boundary between crust and mantle, temperatures range from 500 to 1000°C (932 to 1864°F). At the deeper levels of the mantle bordering on the outer core, the temperature rises to 4000°C (7230°F).

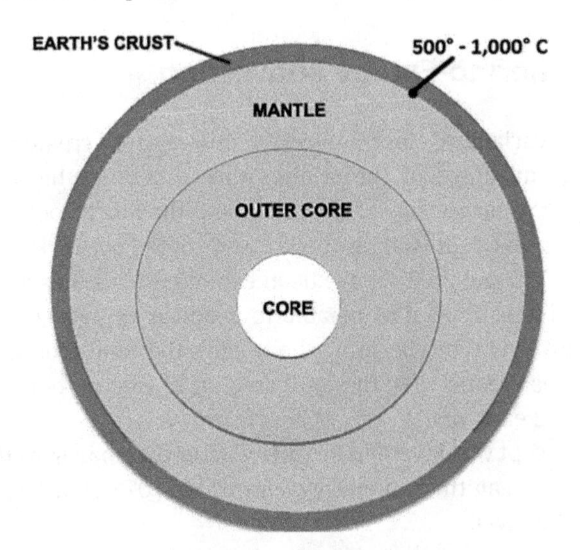

Figure 125 Cross-section of Earth and temperatures.

The deepest we have ever dug is one-third of the distance between surface and mantle. But we may not have to dig at all. There are several ways to reach the mantle without digging through the crust. The crust at the bottom of the oceans is generally much thinner than the continental crust, and there are areas on the

Atlantic seafloor (midway between Cape Verde and the Caribbean Sea) where the mantle lies exposed without any crust covering at all only 3 km beneath the ocean surface and covering an area thousands of square kilometers.

We know temperature strongly influences seawater conductivity.[122] By doping columns of saltwater extending from the 1000°C mantle up to the 15°C surface of the ocean, the difference between those two temperature levels could produce enough energy to meet much of the planet's needs.

Active volcanoes also provide direct access to an almost unlimited source of extreme heat. It needs some careful design work and some clever application of material science, but the potential is there to create electrical current from every one of Earth's active volcanoes. Materials such as tantalum carbide (TaC) and hafnium carbide (HfC) have already been developed that can withstand temperatures of 4000°C.

Ocean Thermo-Energy Conversion

This is a variant of thermoelectric power that creates energy by taking advantage of the temperature difference between the relatively warmer surface of the ocean and the much cooler waters existing at great depths at particular parts of the ocean. Differences between 9.7°C and 28.0°C have been exploited. This meager 18.3°C differential is estimated to be able to produce up to 88,000 TWh/year of power without harming the ocean's thermal structure. That is four times greater than the total amount of electricity consumed by the world each year.

The concept was invented in 1881. Currently, Japan runs the only operational ocean thermo-energy conversion (OTEC) facility on the island of Okinawa.

[122]T. M. Dauphinee and H. P. Klein, The effect of temperature on the electrical conductivity of seawater, *Deep Sea Research Part II Topical Studies in Oceanography*, **24**, 1977, 891–902. See also: A. M. Assiry, M. H. Gaily, M. Alsamee, and A. Sarifudin, Electrical conductivity of seawater during ohmic heating, *Desalination*, **260**, 2010, 9–17.

Figure 126 Ocean thermo-energy conversion on Okinawa, Japan.

Hydrogen

Hydrogen is considered a potentially pollution-free fuel for the future because:

1. It does not contain carbon,
2. It has the maximum energy content per unit mass of all the known fuels, and
3. The only by-product of its combustion (burning) is water.

In nature, hydrogen is almost always found as part of another compound, such as the vast reserves of hydrogen in the Earth's water. Because high temperature steam has been required to isolate it, 95% of the world's hydrogen is made from the burning of fossil fuels.

There are several technologies being developed to separate hydrogen for use in vehicle fuels.

The challenge is to produce it without needing to combust fossil fuels.

Superconductivity and the Electron Generator

This could be both a potential source of power and a potential way to store alternative energy. As a power source, it has greater velocity and larger potential application than atomic energy with none of its negative by- products. As an energy storage device, it has literally unlimited potential. It has been called the electron generator, and scientists have skirted around its edges for over 100 years since the Dutch physicist H. K. Onnes first observed superconductivity in 1911. It has never got off the ground because scientists never oriented it against elementary physics.

When certain metals approach Kelvin zero or a near zero (−273°C), their conductivity becomes almost infinite while their resistance approaches zero and magnetic properties are nullified. At that temperature, there is close to no motion in the object and you can pump electricity into it. In fact, because it has no resistance, it has close to infinite capacity. Hubbard posited that if you get these metals down to as close to −273°C as possible, you can keep pumping electricity into them until they contain a very large charge—maybe billions of megavolts. When used for storage, it would have almost infinite capacity.

At −273°C, aluminum is almost in a powder form, if it did not have a crust. Any of our metals when brought down to this terrifically low level gets into one of these states where it has an infinite capacity.

Now you keep that package of charge near −273°C, and when you want to release some of its energy, you just heat it up a little. Just a little. It is not much of a job to do that, and the second it starts to heat up, it starts releasing electricity. This is an enormously powerful force, releasable in a very short space of time, and containable in a very tiny cubic space.

According to studies by Cornell University and Brookhaven National Laboratory, it is the presence of **ridges** that suppresses the use of superconductivity. When they looked at superconducting materials with a scanning tunneling microscope that can scan a surface in steps smaller than the width of an atom by reading the energy levels of electrons, they found the electrons formed ridges (alternating rows of many or fewer electrons), "like the alternating compression of air molecules in sound waves." They thought these ridges were an anomaly, not realizing they were actually one of the three basic forms of energy, and what they were looking at was a condensation/rarefaction process.

Armed with that knowledge, someone could use this mechanism to make a breakthrough in 21st-century energy production.

Nikola Tesla's Free Energy

When Albert Einstein was asked by a reporter
how it felt to be the smartest man on Earth, Einstein replied:
"I wouldn't know. Ask Nikola Tesla."

Tesla's biographers delight in pointing out some of his more peculiar personal traits, but they are irrelevant. From childhood he had the goal *"to harness the energies of nature to the service of man"* and he successfully pursued that goal his entire life.

Tesla's polyphase alternating current inventions changed the world. The system could transmit and distribute power over great distances via high-voltage lines, making electricity both cheaper and far more accessible. His polyphase generators could be built for much higher power levels and rotating speeds than were previously possible, at less cost and required less maintenance.

Figure 127 Nikola Tesla demonstrating his control of energy.

The Tesla coil produces high voltages and high-frequency currents using the resonance of two or more oscillating circuits.

He designed and patented lights in which illumination is created through the mechanism of arcing between two electrodes. He patented lamps powered by high-voltage and high-frequency current in which one or both electrodes are placed within a vacuum. He discovered he could also create illumination from high-frequency current put through diluted gases, and although those fluorescent tubes are now relics of the past, the importance of that discovery is relevant to Energy Miracles—what he found was that high-frequency current causes ionization of the gas in the tubes. He believed it was possible to illuminate the whole world by ionizing the gases of the upper layers of the atmosphere with high-frequency currents.

Only someone with a deep understanding of energy and electricity could have ever envisioned such systems, much less brought them into existence. At the time, nobody could match his understanding and control of electricity. In his New York laboratory, he would demonstrate the "harmlessness" of alternating current by manipulating glowing orbs of pure electrical energy with his bare hands, sometimes passing them to his friend Mark Twain. Nightly at the Chicago's World Fair in 1893, Tesla turned himself into a human conductor, linking the huge electrical turbines to a motor by passing the electricity through his body.

But his most important work was in the wireless transmission of radio signals and electric power. He discovered that if you sent a radio wave into water, it would be transmitted in all directions; in order to receive the signal, all you need do is insert a wire into the water and "listen" to it. The same was true of a signal sent through the Earth. And what he had his sights set on as early as 1899 was exactly that. He planned to cheaply send out massive amounts of electricity and floods of data signals to every location on the planet, by using the core of the Earth as his conducting medium. His principal idea was to send electromagnetic waves by a sufficiently powerful transmitter to spread completely over the Earth's surface. When the waves reached the opposite side, they would reflect and combine with the initially transmitted waves, creating standing waves. A user anywhere in the world would be able to draw energy from those standing waves through an inexpensive receiver.

Tesla went to Colorado Springs between 1899 and 1900 to pursue this idea. He had a friend, a director of the Colorado Springs electric company, who had offered him unlimited power for the experiments he had in mind. Tesla built a tower outside of town and went to work. By that time, many experimenters had wirelessly sent small packets of energy out to small distances. The most successful of them, Marconi, had transmitted radio waves out to a distance of 2 miles. In this regard, Tesla was way ahead, having already sent radio waves out to a distance of 30 miles. He had also demonstrated a radio-controlled boat in Madison Square Garden in 1897. Control of the direction, speed, and lights was accomplished with the "tuning" of radio signal frequencies. Tesla knew more about this process than anyone on Earth and in fact obtained the first patents on it. But he was not interested in transmitting radio waves. He thought they were wasteful. **He was interested in long-distance wireless distribution of electricity.**

Using the core of the Earth as a conductor for large electrical currents presented different challenges than sending radio signals through the atmosphere. A radio signal sent through the air also goes out in all directions. With the right equipment, a message sent from New York to Washington can be overheard in London. But because the particles in the air are so much less dense than those in the more substantial matter of water or the Earth's core, the frequency used for the signal becomes paramount. Think of these frequencies as different pathways operating simultaneously in the same space.

And he was in Colorado Springs with this bigger goal in mind. The huge radio tower he constructed was not for broadcasting radio waves. It was for producing electricity.

Figure 128 Tesla laboratory in Colorado Springs.

There is a weather phenomenon called St. Elmo's fire that occurs in the presence of strong electrical fields such as those in lightning storms. A halo-like glow appears around pointed objects such as the masts of ships. Once Tesla's tower was up and running, the entire neighborhood glowed with St. Elmo's fire every night. Not only that, the city's fire hydrants created an electrical arc if a metal object was brought close to one. What Tesla was doing was pumping electricity directly into the Earth. He was developing what he called a mechanical oscillator, an electric generator with a self-contained power plant that did not rely on an external steam engine. In other words, he was inventing a steam turbine without the steam. He viewed his invention as a way to use the planet itself to propagate an electrical current of enormous strength. With his vast electrical signal, Tesla would water the deserts, alter the climate, illuminate the sea-lanes at night, and provide everybody in the world with unlimited quantities of free electricity.

With radio waves, one is dealing with frequencies of tens of thousands to millions of cycles per second. Transmitting electrical power is an animal of a different color, described in Tesla's own words as follows:

"The practical applications of this revolutionary principle have only begun. So far, they have been confined to the use of oscillations that are quickly damped out in their passage through the medium. Still, even this has commanded universal attention. What will be achieved by waves that do not diminish with distance baffles comprehension.

"It is difficult for a layman to grasp how an electric current can be propagated to distances of thousands of miles without diminution of intention. But it is simple after all. Distance is only a relative conception, a reflection in the mind of physical limitation. A view of electrical phenomena must be free of this delusive impression. However surprising, it is a fact that a sphere of the size of a little marble offers a greater impediment to the passage of a current than the whole Earth. Every experiment, then, which can be performed with such a small sphere can likewise be carried out, and much more perfectly, with the immense globe on which we live. This is not merely a theory, but a truth established in numerous and carefully conducted experiments. When the Earth is struck mechanically, as is the case in some powerful terrestrial upheaval, it vibrates like a bell, its period being measured in hours. When it is struck electrically, the charge oscillates, approximately, twelve times a

second. By impressing upon it current waves of certain lengths, definitely related to its diameter, the globe is thrown into resonant vibration like a wire, stationary waves forming, the nodal and ventral regions of which can be located with mathematical precision. Owing to this fact and the spheroidal shape of the Earth, numerous geodetical and other data—very accurate and of the greatest scientific and practical value—can be readily secured... By proper use of such disturbances a wave may be made to travel over the Earth's surface with any velocity desired, and an electrical effect produced at any spot which can be selected at will and the geographical position of which can be closely ascertained from simple rules of trigonometry.

"This mode of conveying electrical energy to a distance is not 'wireless' in the popular sense, but a transmission through a conductor, and one which is incomparably more perfect than any artificial one. All impediments of conduction arise from confinement of the electric and magnetic fluxes to narrow channels. The globe is free of such cramping and hinderment. It is an ideal conductor because of its immensity, isolation in space, and geometrical form. Its singleness is only an apparent limitation, for by impressing upon it numerous non-interfering vibrations, the flow of energy may be directed through any number of paths which, though bodily connected, are yet perfectly distinct and separate like ever so many cables. Any apparatus, then, which can be operated through one or more wires, at distances obviously limited, can likewise be worked without artificial conductors, and with the same facility and precision, at distances without limit other than that imposed by the physical dimensions of the globe."

His idea was elegant. He embedded electrodes into the Earth, and then fired an electrical impulse into them. Initially he would be limited by the power made available by the local electrical company. But when the impulse returned from the other side of the world, he believed he would be able to fire it back again, propagating it through the Earth's core. Furthermore, he computed that hydropower from existing water resources would be enough to prime his system worldwide.

"But we shall not satisfy ourselves simply with improving steam and explosive engines or inventing new batteries; we have something much better to work for, a greater task to fulfill. We have to evolve means for obtaining energy from stores which are forever inexhaustible, to perfect methods which do not imply consumption and waste of any

material whatever. Upon this great possibility, which I have long ago recognized, upon this great problem, the practical solution of which means so much for humanity, I have myself concentrated my efforts since a number of years, and a few happy ideas which came to me have inspired me to attempt the most difficult, and given me strength and courage in adversity...I have made progress, and have passed the stage of mere conviction such as is derived from a diligent study of known facts, conclusions and calculations. I now feel sure that the realization of that idea is not far off; namely, that of the operation of engines on any point of the Earth by the energy of the medium."

With a simple device, easily manufactured, he conceived consumers would be able to power their houses with limitless electricity. Once his tower and equipment were set up, he was ready for his test. Wearing thick rubber-soled shoes, he placed himself at some distance from the equipment. At his signal, an assistant turned on the dynamo. A 6-foot bolt of electricity flew from the top of the tower. (The accompanying thunder was audible 22 miles away.) As the test continued, the length of the lightning bolts increased in strength and length up to 130 feet. At that point, the tower and the entire city of Colorado Springs went black—the local power station had burned out.

Undaunted, he went back to New York convinced he was on the verge of discovering the most powerful force in the history of the human race. With a loan from J. P. Morgan, he bought a tract of land called Wardenclyffe on Long Island and began to build a new, perfected, 187-foot electrical tower (Fig. 129).

Again, he made it clear he was not planning to use radio waves because they were too weak and inefficient. His few published papers described what he intended to do but did not reveal his method. We know his Tesla coil was an immensely powerful device for magnifying the power of electricity, and he claimed the "Tesla currents" would be 90% efficient. Lord Kelvin, another electrical genius, came and looked and listened to Tesla's explanations and emerged from the meeting a staunch supporter.

Tesla's investors were heavily invested in power plants and had no intention of supporting the idea of "free electricity for all." They told Tesla that the money they had given him was to be used only to wirelessly broadcast yachting and horse racing results together with stock quotations. When Tesla refused, his sources of funding evaporated overnight. Tesla retreated, and the Wardenclyffe Tower was dynamited for scrap in 1917.

Figure 129 Wardenclyffe on Long Island, New York.

Tesla's claims were not those of a madman. He was one of the last who was able to turn energy dreams into actuality. You can see operating examples of some of his devices at Tesla's Birthplace Museum, at the site of Tesla's childhood home in Smiljan, Croatia. A much larger and more comprehensive selection of his equipment is to be found in the Nikola Tesla Technical Museum, in Zagreb, capital of Croatia.

His work has particular application to Energy Miracles. One of the fastest ways to concentrate and get electricity is to shoot it through copper. That is why copper is used so predominantly in generators and electrical distribution. Tesla was the first to realize there is no reason why we could not have the same generation operating without a single wire. All you would have to have is an intensity and wavelength of flow that could travel through air or through the medium of the Earth. Tesla was able to do that, not just with radio waves, but with electricity as well. He was able to turn light bulbs on wirelessly at a distance of 30 miles. Unfortunately, he died before this dream was fully realized. But it is a line of research that could be picked up and followed that could lead to an Energy Miracle.

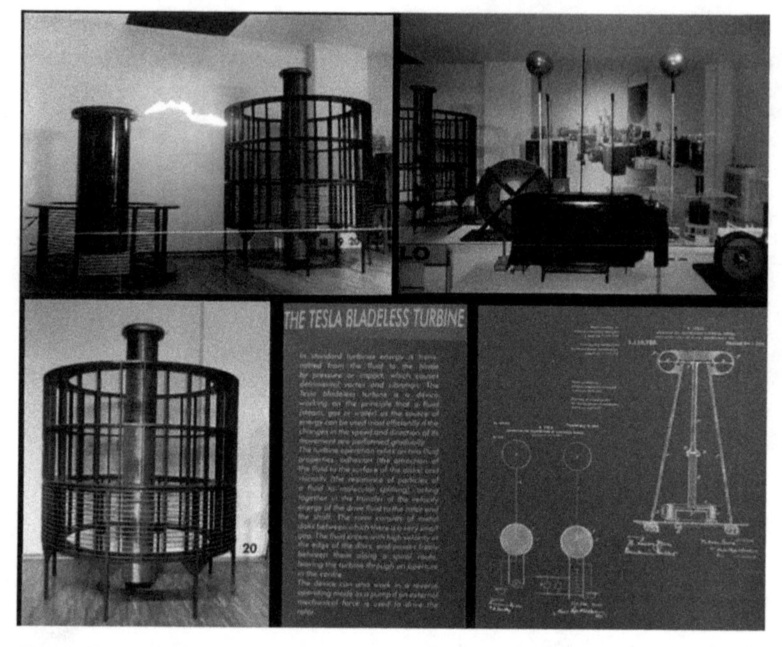

Figure 130 Nikola Tesla Technical Museum in Zagreb.

Many of Tesla's drawings and papers are available. Some are accessible on the energymiracles.net website. But 30 barrels and bundles were confiscated by the U.S. government in 1934, when, in a moment of paranoia, it feared his work could potentially be used as a weapon. Aside from those 30 barrels, immediately upon his death on January 7, 1943, the Alien Property Section of the U.S. Department of Justice moved into Tesla's lab and apartment and confiscated all of his scientific research, including his work on Wardenclyffe and research on the ionosphere. This amounted to 80 trunks of materials and his experimental machines (some of which were operational). It took two truckloads to remove them to the Manhattan Storage and Warehouse Company in New York City, the same facility where the materials confiscated in 1934 were being housed. The Federal agents photographed everything, and the Justice Department asserted authority over all of them. The Justice Department has replied to the numerous requests to release these materials that they "do not know where they are."

The United Nations should request all his materials be located and immediately released and made public for the benefit of all humankind. (The branch of the Justice Department that claimed authority over these materials—the Alien Property Section—in fact probably did not have that authority since Tesla was a naturalized U.S. Citizen, not an alien.)[123]

Re-tasking Hydropower

Of the existing renewable energy sources, hydroelectric power from dams trumps all. The world's nine largest operating power plants of any source are all hydroelectric plants. Four of those nine are in China (including the largest by far at the Three Gorges in Sichuan). Three are in South America (including the overall number 2, the Itaipu Dam (on the border of Brazil and Paraguay)). The U.S. has the Grand Coulee, and number 9 is in the Republic of Khakassia in Russia.

As was discussed earlier in the section on Tesla currents, the key to his "free energy" is producing an exactly correct type of wave, of a precise frequency and wavelength. He needed some kind of conventional power plant to energize his system. He tried the power plant in Colorado Springs, but it was insufficient. He was confident that the existing hydroelectric resources at the time would be sufficient to run his global electrical network, even though only one of the preceding listed nine huge hydro-plants was in operation when Tesla died (Grand Coulee).

This means if Tesla's free energy should be rediscovered (or some other type of potential electrical source that needs to be amped up to get running), existing hydroelectric dams could be a potential way to do it, since they release no emissions, are renewable, reliable, and cheap, and can be run at large scales.

They could be gradually re-tasked to power up the new free-energy systems.

---

[123]More data on Tesla, including original Tesla materials, and correspondence with the U.S. Department of Justice, can be downloaded from the energymiracles.net website.

Magnetism

In the 1980s at Virginia Tech, Gerhard Beyer and Howard Johnson did groundbreaking work on the magnetic fields that surround magnets. They found that magnets were able to fire bursts of energy like guns, and these forces can be up to a thousand times more powerful than the base magnet strength. They were able to demonstrate a small magnet could power a small car indefinitely—the car's bearings would wear out before the magnet. The functions of magnetic fields are far more complex than what we have been taught, and the potentials are limitless considering the fact that the Earth itself is a giant magnet.

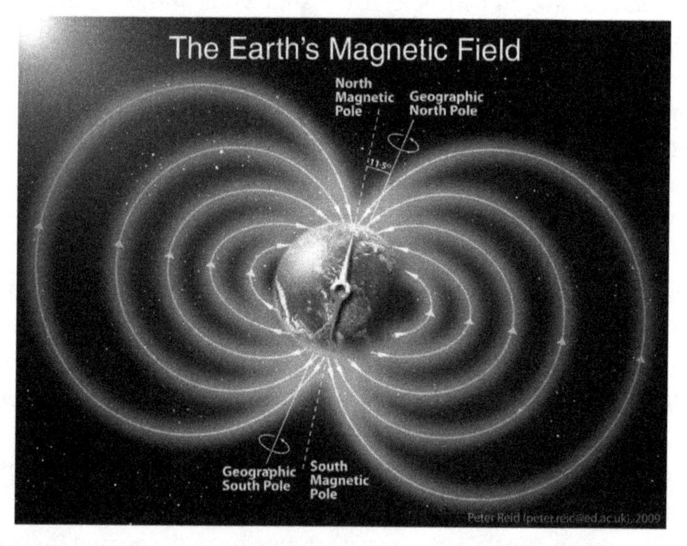

Figure 131 Earth's moving magnetic field. Sometimes the field points toward the center; sometimes it points away from it.

Nuclear Fission and Fusion

Fission and fusion are two ways of altering atoms to create energy. Fusion joins two lighter atoms into a heavier one, while fission splits a heavier atom into two lighter ones. Fusion has a potential of many times that of fission, but has several regrettable problems associated with it, including energy radiation damage to the reactors and connected structures, radioactive waste, the need for shielding humans who come in contact with it, and the potential for production of weapons-grade plutonium 239, which would add to the threat of nuclear proliferation. It is fission that is used today in all commercial nuclear power plants and is, itself, millions of times more powerful than chemical reactions such as burning coal. The ferocity of the explosions involved requires both processes to be controlled in nuclear reactors. The high costs, safety concerns, and dangerous waste products of nuclear power plants would seem to put it out of the running for an Energy Miracle, but a technology that could eliminate all three of those disadvantages would be enormous. A company started by Bill Gates (TerraPower) is working on such a solution. According to John Gilleland, its Chief Technical Officer, "Bill Gates and his colleagues looked at solar, wind, any kind of energy you can think of, and determined that they're all important, and they all have their role, however, nuclear was the only source of energy which could provide the necessary huge quantities that we need on a global basis." Its product is called a Traveling Wave Reactor (TWR®), which is said to be cheaper and safer plus produces considerably less nuclear waste than existing nuclear power plants. It is a work in progress.

Alternative Fuels

Electric battery vehicles are a solution for passenger cars, buses, and trucks that travel shorter distances. It is not practical to have electric batteries power airplanes, cargo ships, and the 18 wheelers that transport so much of our freight because the batteries would have to be too big. For the long-haul applications, some kind of cheap alternative fuel is needed. Ethanol was the first such fuel, but now there are many more, made from plants that are not grown for food to avoid fertilizers, which are big carbon emitters.

Solar and Other Zero-Carbon Fuels

Notwithstanding limitations of solar power from solar panels discussed earlier, the sun provides us with a potentially unlimited source of power. The energy hitting the Earth from the sun in just 1 hour exceeds the amount of energy used on all seven continents for a whole year. If we could just harness a small fraction of this...

Turning sunlight into chemical energy is what plants do every day. It is called photosynthesis, and scientists could adapt it to a larger purpose. One company is using sunlight to split off the hydrogen atoms of water and then combine them with carbon dioxide to produce a solar fuel-source directly from the sun. It is another work in progress because thus far the cost of doing it makes it impractical for large-scale application. To make it cheaper, they would need to create or discover new materials.

Another related idea is to cover large areas of desert with artificial turf made of plastic cells able to capture sunlight to make fuel. Each cell would contain water and a catalyst capable of accelerating the chemical reactions necessary to produce hydrogen or carbon-based fuels. Choosing a workable catalyst has thus far been problematic because the best ones (like platinum) are also the most expensive. But this is an obstacle that can be overcome. There are folks working on producing zero-carbon fuels with wind and solar power. The fuels so produced can be turned back into electricity.

Carbon Management

Carbon management is a suite of technologies with the purpose of significantly reducing or eliminating greenhouse gas emissions from fossil fuel power plants and industrial facilities. These include CCUS (carbon capture, use, and storage) and CDR (carbon dioxide removal). CDR can pull CO_2 directly out of the air. Carbon direct air capture is not a new idea. It has been used in submarines for many years. But it has so far been prohibitively expensive for large-scale deployment. An advantage of all carbon management technologies is that CO_2 can be safely stored underground or turned into useful products.

Where coal is being used, it is because it is cheap and available. Burning "clean coal" has been a dream for 50 years, and technologies exist for accomplishing this to some degree, but have never been deployed because they would double or triple the cost of using coal. This would only be a stopgap measure until a real Energy Miracle is created, but it could be beneficial to society in the meantime.

Energy Storage

We need reliable and widely usable ways to store renewable energy sources for days, weeks, or months. Thermal-powered storage technologies have the potential to afford such storage. One such system operates like a heat pump and stores heat from renewable sources in molten salt. In discharge mode, it produces electricity.

Rescue the Terrified Geniuses Hiding in Caves

I know this sounds weird, but there are people who have started on the quest for an Energy Miracle, made substantial progress, and then became fearful for their lives if they ever broadcast their results. It does not matter whether this was paranoia on their part, or the threat was actual. By broadcasting an international request for Energy Miracles, their safety can be assured, and they may be encouraged to come forward. Over the course of the years, I have heard of three such cases myself. They really exist.

Annexure

Lightning's Downward Explosive Discharge: A New Model

Lightning is the greatest showcase of natural energy. It is everywhere, and it is big. For the past 80 years, lightning research has suffered much the same fate as research into new methods of energy production. And for the same reasons. This section on lightning is a relatively nontechnical expansion of a paper written for the 2021 joint session of the International Conference on Lightning Protection (ICLP) and International Symposium on Lighting Protection (SIPDA), a conference sponsored by the IEEE.

Figure 132 Demons do not cause lightning.

Where does lightning come from? For many centuries in Europe, agreement was near unanimous on the answer to that question and that agreement transcended the usual divisiveness of European countries and their religions. Brits and French, Italians and Spanish, Protestants and Catholics were all in accord: lightning was caused by Demons. And the solution to it was to ring the church bells.

Unfortunately, this theory was just one more example of theory in conflict with observable fact. Within a span of just 33 years in Germany alone, lightning damaged 400 church towers and caused 120 bell-ringers to be electrocuted, crushed by debris, blown to atoms, or otherwise sent on to the next life.[124]

Current Lightning Model

Ideas about lightning remained nebulous until Benjamin Franklin took up the subject as part of a broader study of electricity in the 1750s. It was Franklin who proposed that electricity was actually a single force acting between two different potentials and that lightning was electrical charge. These incisive discoveries sparked a flood of progress in both electricity innovation and lightning research. It can be argued that forward progress in both fields became severely impeded by the mid-1930s, and many currently accepted theories of lightning are not much more accurate than the Demon model. Lightning experts Martin Uman and Joseph Dwyer published a comprehensive paper on lightning physics in 2014 in which they remarked the irony that we could know so much about things half a universe away while not understanding the process of lightning going on just a few miles above our heads. There is a reason for that, and it is called the *fixed idea*. At various times in history, scientific progress became blocked by a *fixed idea*—a strongly held belief that may have looked good on paper but did not really add up. Looking for such a culprit in the subject of lightning science brought to view a long-standing misconception about one of lightning's favorite actors: **the return stroke**. Other misconceptions were found, as well.

We shall suggest a new simpler lightning model, based on actual observations and basic physics.

First a quick review of the currently accepted lightning model:

1. A preliminary charge breakdown somewhere in the storm cloud creates a high-voltage potential with respect to the ground. (It is believed this is related to the buildup, motion, and collision of ice crystals within the clouds, but we have no conclusive understanding of the process.)

[124] L. J. Davis, *Fleet Fire*, Arcade Publishing, New York, 2003.

2. Negative charge is lowered to the ground in a series of steps. (This is called the stepped-leader.) As the downward leader approaches the ground, an electrostatic process induces positively charged streamers to be emitted upward from the Earth. When the downward moving stepped-leader contacts one of those upward-moving discharges (some tens of meters above the ground), the leader tip becomes connected to ground potential.
3. The leader channel is then discharged when the highly luminous first return stroke propagates continuously back up the previously ionized and charged leader path. (This is called the return stroke.)

Lightning occurs all over the world, millions of times each day. There is hardly a single person on all the seven continents who has not personally experienced it, including you who are reading this. Ask any of them whether the force of lightning goes from the sky to the ground or from the ground to the sky and he will either look at you a little strangely, like you are crazy, or a little slyly and wonder if you are throwing him a trick question. You assure him you are serious, and he will say, "Of course it goes from the sky to the ground." Ask a lightning scientist, and the answer would not be so simple. He will tell you about positive and negative cloud-to-ground lightning, return strokes, and about upward lightning. The most recent international lightning standard proclaims: Two basic types of flashes exist: downward flashes initiated by a downward leader from cloud to Earth; upward flashes initiated by an upward leader from an earthed structure to cloud.[125]

Lightning scientists across the world agree that 90% of the lightning that impacts the Earth's surface is of the negative cloud-to-ground (CG) variety. The preposition "to" in the phrase "cloud **to** ground" assumedly means "in the direction toward." So far, so good. But now we must introduce a new term: the "**return stroke**." Nothing so clearly demonstrates the quantum confusion surrounding the subject of lightning than the various attempts made to define return stroke. Look it up in textbooks or Google the term, and you will find the following definitions:

---

[125]IEC 62305-1 Annex A

1. "The main discharge in a lightning flash" (i.e., charge without direction);
2. "The very bright flash that we see as lightning" (i.e., no charge, just illumination);
3. "The movement of the charge starts at the point of contact with the ground and rapidly works its way upward" (i.e., charge that moves upward);
4. "The visible flash is associated with the rapid movement of charge downward" (i.e., visibility associated with charge moving downward);
5. "The principal discharge in a lightning stroke carrying the main current upward from the ground to the cloud" (i.e., charge and current moving upward);

So, which one is it? You may be surprised to learn that the universal agreement of experts is "5": The main discharge of lightning current travels from the point of contact with the ground, back up to the cloud. But this is a rather recent theory.

Understanding lightning requires an understanding of energy and light. Before 1750, it was universally held that light was composed of small particles traveling in straight lines and that electricity was an interaction between two totally different kinds of electrical forces. Ignoring almost the entire scientific community of the time, Benjamin Franklin proposed: (1) that light propagated in a medium through waves; (2) that electricity was actually a single force between two different potentials; and (3) that lightning was electrical charge. These incisive shifts in viewpoint sparked a flood of progress and innovation in electricity research. That progress continued until about 1938, when it abruptly ceased.

Lightning is the most available manifestation of natural energy generally available to a researcher. For that reason, the subject is included in this book. Although no other natural phenomenon has received as much examination in the past 100 years, the same questions about charge separation, propagation, and attachment that tormented scientists a century ago remain unsolved today. There is a reason why lightning has escaped better understanding in the face of nearly a century of intense scrutiny by so many competent minds.

Although this section may not immediately answer every question there is about lightning, it is likely that using some of the observations made here will enable researchers to persuade lightning to disclose more of its better-kept secrets.

The Return Stroke

In their encyclopedic "*Lightning: Physics and Effects*," Uman and Rakov labeled the return stroke "the most studied lightning process."[126] Since it is generally accepted to be at the center of the lightning phenomenon, this is the first thing that will be addressed here. Contemporary books and articles about lightning begin any description of the return stroke by reminding us of the stepped-leader that comes before it, creating a conductive path between the cloud and ground. This will invariably be followed by a statement like: "The return stroke traverses the leader path upward from ground to the cloud charge source and neutralizes the negative leader charge."[127]

Lightning has not always been thus described. Before the 1930s, the term "return stroke" never appeared in physics texts. In the first quarter of the 20th century, lightning was understood to be simply "an electric spark of great magnitude... between a cloud and the Earth."[128,129]

When lightning was turned upside down

The modern definition of "return stroke" was conceived and popularized in the 1930s by an American electrical engineer named Karl B. McEachron. McEachron, at the time working for GE (General Electric Corporation), was studying the lightning discharge by recording lightning strikes to the Empire State Building using a so-called "streak camera" designed by Sir Charles Boys.

---

[126]V. A. Rakov and M. A. Uman, *Lightning: Physics and Effects*, Cambridge University Press, Cambridge, UK, 2003, p.143.

[127]Y. Baba and V. Rakov, Present understanding of the lightning return stroke, in *Lightning Principles, Instruments and Applications*, Betz, H. D. (ed.), Springer Science and Business Media B.V., 2009.

[128]J. A. Culler, *A Text Book of General Physics: Electricity, Electromagnetic Waves, and Sound*, J. B. Lippincott Company, Philadelphia, 1914, p.272.

[129]G. C. Simpson, On lightning, *Proceedings of the Royal Society of London A*, 1926.

McEachron had been an instructor in electrical engineering at Ohio Northern and Purdue Universities between 1914 and 1922, after which time he went to GE to became head of the research and development section of the lightning arrestor department. His boss at GE was the brilliant German physicist and inventor Karl Steinmetz. Steinmetz had made his greatest contributions in the 1880s and 1890s, including revolutionizing AC circuit theory and analysis by substituting a simple algebraic process for the complicated calculus-based methods previously in use. This formula is still used in the design and testing of motors. At GE, Steinmetz was a pioneer of human-made lightning, being the first to create artificial lightning in his football field-sized laboratory and high towers. He was called *the forger of thunderbolts.*

By the time he met McEachron in 1922, Steinmetz had become badly enmeshed in quantum entropy and Einstein's relativity.[130] For over a year, until he passed away in 1923, Steinmetz was McEachron's boss and mentor, and it is clear that Steinmetz's quantum confusions rubbed off on the younger McEachron. Just a few years later, McEachron began claiming that the most damaging type of lightning discharge was the upward stroke from ground to cloud.[131] Acceptance of McEachron's work was far from universal. His contemporaries showed little respect for his upside-down lightning. Bruce and Golde published an article critical of McEachron's methods and results.[132] Golde refused to include any McEachron references in his comprehensive book, *Lightning*.[133]

Malan, though more tactful, would still not cite McEachron's works in his books and articles.[134] Loeb had this to say of him, "His

---

[130]See for example Steinmetz's 1912 article "The death of energy and the second law of thermodynamics" and his "Four Lectures on Relativity and Space," given in the last year of his life in 1923 while mentoring McEachron.

[131]*The Encyclopedia of Earth*, article: McEachron, Karl B., Author: IEEE, Creative Commons Corp 2007.

[132]C. E. R. Bruce and R. H. Golde, The lightning discharge, *Journal of Electrical Engineers*, **88**, 6, 1941.

[133]R. H. Golde, *Lightning*, Academic Press, London, 1977.

[134]D. J. Malan, *Physics of Lightning*, English Universities Press, London, 1963; D. J. Malan and B. F. J. Schonland, The electrical processes in the intervals between the strokes of a lightning discharge, *Proceedings of the Royal Society of London A*, **206**, 1085, 1951, 145–163; D. J. Malan and B. F. J. Schonland, Directly-correlated photographic and electrical studies of lightning from near thunderstorms, *Proceedings of the Royal Society of London A*, **191**, 1027, 1947, 485–503.

data were not well controlled and will not be discussed further."[135]

Undaunted, McEachron produced a movie to popularize his "return stroke" theory, after which he was to become known as "The Thunderbolt Hunter." McEachron's movie (released by GE about 1938) did not pull any punches: "Contrary to what we've always thought, the really destructive discharge goes from Earth to cloud... Thus, scientific research forces us to revise some of our pet notions of what happens, when and how, during a lightning stroke...."[136] Shades of quantum logic.

Although there was no credible evidence to support his claims, the movie was intriguing enough to upset the certainty that lightning current went from cloud to ground, and ever since 1938, we have had upside-down lightning. In hindsight, it has always been easy to prove the actual direction of the lightning charge. Experts unanimously agree that up to 90% of all lightning never hits the Earth; it discharges up in the clouds from one spot to another spot in the same cloud or from one cloud to a neighboring cloud. So, when electrical charge builds up in the clouds, there is only a 10% chance that it will discharge on the Earth's surface. (See Fig. 133.) It is logical this would be the case since to hit the ground, lightning must overcome the resistance of the air between the cloud and the ground. The distance between neighboring clouds is much shorter than that of the clouds to ground. Less distance means less resistance, which makes it easier for the lightning charge to dissipate up in the clouds.

Ten percent goes from cloud to ground. There is no ground-to-ground lightning.

Now, if charge originated at ground level and flowed from the ground **up** to the clouds, it follows that there should also be some, if not many, "ground-to-ground" lightning strokes where lightning originating at ground level discharged itself at the much closer and thus more attractive locations at ground level. A high percentage of ground-to-ground lightning would prove the contention that lightning propagates upward from the ground to the clouds. Even a single case would be interesting. But of course, no such ground-to-ground lightning has ever been reported.

---

[135]L. B. Loeb, Contributions to the mechanisms of the lightning stroke, *Monthly Weather Review*, **95**, 12, 1967, p.828.

[136]*Thunderbolt Hunters: An Excursion in Science*, produced and distributed by General Electric Corporation, circa 1938.

Figure 133 Ninety percent of lightning discharges within the clouds.

Up or down? The power of optics

Does a cloud-to-ground lightning stroke propagate upward or downward? This is not a trick question, and though it may sound provocative to those who have already given this much thought, this question generated enough interest to prompt the eminent Prof. Martin Uman to release an article entitled "Does a stroke between cloud and ground travel upwards or downwards?"[137]

McEachron may have been somewhat led astray by the optical results of the camera he used. Figure 134 is a photo taken with a

[137]Distributed in 1998 at an exhibition in Tampa, Fla. entitled "Petrified lightning from Central Florida." Article originally published in M. A. Uman, *All About Lightning*, Dover Publications, New York, 1986.

streak camera similar to the one used by McEachron.[138] A streak camera employs two separate lenses focused on a single moving filmstrip from which it is possible to visually deduce the direction and speed of the luminous parts of the lightning process. Line A-B (Fig. 134) shows the downward moving stepped-leader—luminous, but just barely. Line B-C shows the upward-moving highly luminous wave that has been universally called the return stroke. The first and brightest flash appears at the bottom of the channel at Point B. The last and least luminous spot is at the top at Point C.

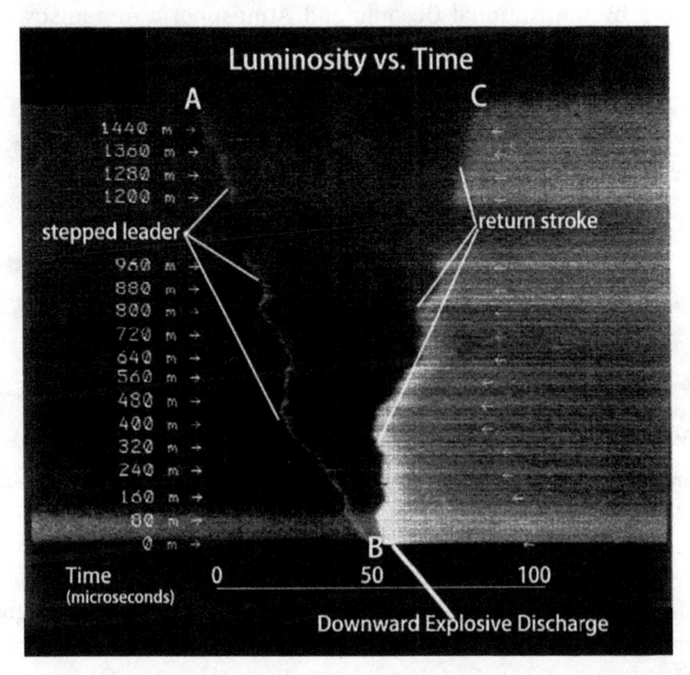

Figure 134 Streak camera photo of a cloud-to-ground lightning stroke (abstracted from Jordon et al.).

Figure 134 could make it appear that the lighting return stroke propagates upward, but the photo itself is not definitive. The streak camera's limitation is that it can only produce a picture of luminosity over time. As Schonland explains: "The photographic method of studying the lightning discharge by means of the Boys camera has the unique advantage of giving direct information concerning events

[138]D. M. Jordan, V. A. Rakov, W. H. Beasley, and M. A. Uman, Luminosity characteristics of dart leaders and return strokes in natural lightning, *Journal of Geographical Research*, **102**, 8, 1997, 22025–22032.

in the discharge in two dimensions of space and one of time... **The luminous events which it records are, however, secondary processes, and the primary movements of electrical charge which cause them can only be inferred.**"[139]

Prof. Martin Uman and Nobel Laureate Richard Feynman have both remarked that optics may play a role in the confusion between "upward or downward" notions of the lightning flash.[140,141] A clear animated demonstration of the charge pouring out the bottom of the lightning channel while the luminosity is seen moving upward was created by the National Oceanic and Atmospheric Administration (NOAA) and can be viewed on its website.[142] See also Fig. 135.

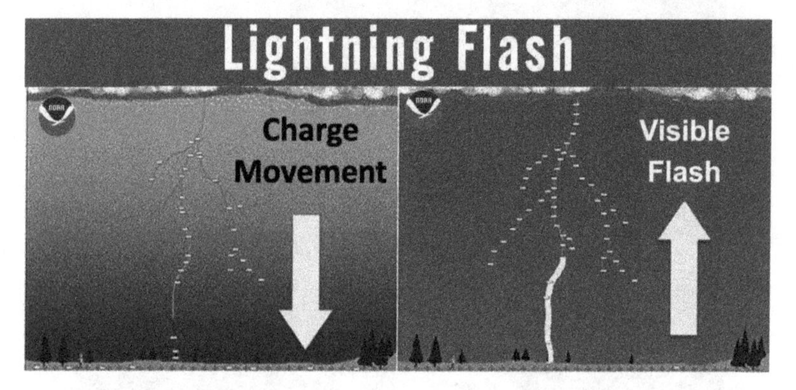

Figure 135 NOAA animation. The actual charge is moving downwards. Only the visual flash propagates upward.

McEachron's streak camera photos showing the luminosity of the flash do not justify his assertion that cloud-to-ground lightning goes from ground to cloud. Neither do subsequently created mathematical models that seek to infer (electrostatically or otherwise) the amount of lightning current that may be traveling back up the return stroke channel when no such lightning current has ever been measured.

[139]B. F. J. Schonland, D. B. Hodges, and H. Collens, Progressive lightning V. A comparison of photographic and electrical studies of the discharge process, *Proceedings of the Royal Society of London A*, **166**, 924, 1938, p.56.

[140]M. A. Uman, *All About Lightning*, Dover Publications, New York, 1986.

[141]F. R. Feynman, "Electricity in the Atmosphere," Lecture 9: The Feynman Lectures on Physics—Lecture 9, Addison-Wesley Publishing Co., Reading, Massachusetts, 1964.

[142]NOAA website animation: http://www.lightningsafety.noaa.gov/science/science_return_stroke.shtml

Karl Berger: what did he say about return strokes?

The science of lightning is fortunate to have its own private encyclopedia, Rakov and Uman's *Lightning: Physics and Effects*, citing over 5000 books and articles and treating every aspect of lightning physics. As for return strokes, the text states: "The most complete characterization of the return stroke in the negative downward flash, the type that normally strikes flat terrain and structures of moderate height shorter than 100m or so, is due to Karl Berger.... The results of Berger et al. (*Electra 41*; 1975) are still used to a large extent as the primary reference source for both lightning protection and lightning research." CIGRE's *Technical Bulletin 549* reaffirmed that exact statement in 2013.[143]

Berger was a Swiss scientist considered the Father of Lightning Research. In his *Electra* 41 article, he analyzed and categorized lightning into its four basic, polarity-based types: He discussed multiple-stroke lightning and continuing currents; he looked at peak currents and waveforms and at flash duration and charge; and he computed mean figures for many of these parameters. But in this landmark article summarizing the results of 32 years of electromagnetic and photographic lightning research, Berger never mentioned the word "return stroke." Not even once.[144]

Figure 136 Mt. San Salvatore (in Switzerland) where Karl Berger spent over 30 years studying, measuring, and recording lightning.

---

[143]CIGRE Technical Bulletin 549, *Lightning Parameters for Engineering Applications*, Paris, 2013.

[144]K. Berger, R. B. Anderson, and H. Kröninger, Parameters of lightning flashes, *CIGRE ELECTRA*, **41**, 1975, 23–37. Available for download at www.energymiracles.net.

The actual origin of return stroke: Third Earl of Stanhope

The term "return stroke" does have an origin. Charles Mahon, Third Earl of Stanhope, a British scientist, statesman, and philosopher, coined it in 1779. Stanhope's book, *Principles of Electricity*, chronicled 71 experiments he conducted into the subjects of electricity and magnetism.[145] In Experiment 16, Stanhope found that an electric current placed in a primary conductor (PC) would inductively magnetize a nearby metallic article (AB). See Fig. 137.

When he placed a second metallic article (EF) end to end with the first (AB) and about 2.5 mm apart, he observed three things: (1) As the voltage increased, sparks would be emitted between the two metallic articles (points B and E); (2) when the primary conductor (PC) was fully charged up and then grounded, it would cause an explosion of force; (3) following that explosion, a single bright spark would then be emitted between points B and E but in the opposite direction.

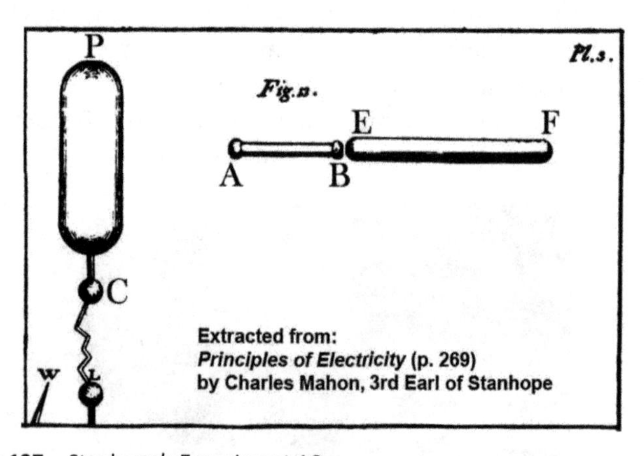

Figure 137 Stanhope's Experiment 16.

On pages 148–149 of his book, Stanhope relates these findings to lightning, distinguishing between what he calls lightning's **Main Stroke**... "in its descent from a Thunder-Cloud to the Earth" and the **Returning Stroke**, by which he referred to the relatively minor event that returned a small portion of that electrical charge back to the cloud.

[145]C. Stanhope, aka C. Mahon, 3rd Earl of Stanhope, *Principles of Electricity*, P. Elmsley in the Strand, London, 1779. Availabe for download at www.energymiracles.net.

Fifty years later, Michael Faraday observed the same phenomenon in his famous "ring," the world's first transformer, just 6 inches in diameter. First establishing a current flow by connecting a battery, he found that when he broke that current flow, a momentary flash of current could be observed in the opposite direction to the main flow.

As far as lightning is concerned, as to evidence for the existence of an actual upward propagating lightning return stroke, Cooray reports, "direct evidence...is not available in the literature."[146]

It is reasonable that there would be *some* disturbance propagating back along the ionized channel already created between cloud and ground. But it would not be very powerful. Any electronic oscillations would be significantly damped. Calling it an "upward propagating return stroke" complicates a basically simple process and interferes with achieving a clear understanding of lightning.

Some real explanations for return strokes

The return stroke phenomenon can be better and more accurately understood in terms of Newton's laws and vacuums. Any energy traveling in space is operating on Newton's laws: for every action there is an equal and opposite reaction. When a lightning strike hits the ground, there is an explosion. The peak power of any explosion is at its vortex, and Krider and Guo confirm that the peak power of a lightning stroke is indeed produced when most of the stroke current is close to the ground.[147] A conservative calculation for the amount of energy in a lightning stroke is 150 billion Joules. That is a serious amount of energy when focused on a single point, but that energy disperses out spherically from the point of impact. See Fig. 138. Most of the charge is spent in the nearby ground, but some will flow back up the already ionized path to the cloud. That minor flow is the return stroke. The cross-sectional area of the ionized path from cloud to ground is only a few centimeters wide. That makes it about 1/20,000th the size of an explosive sphere of 2 m radius. Thus, it can be estimated that only about 1/20,000th the amount of the energy of the lightning strike returns to the cloud in the return stroke. See bottom right frame in Fig. 140.

[146]V. Cooray (Ed), *The Lightning Flash*, Institution of Engineering and Technology, UK, 2003, p.159.

[147]E. P. Krider and C. M. Guo, The peak electromagnetic power radiated by lightning return strokes, *Journal of Geophysical Research*, **88**, 13, 1983, 8471–8474.

Figure 138 Downward explosive discharge (DED); the vast part of the energy goes into the ground and environment.

Another way to explain the return stroke is in terms of vacuums. Students have been taught since Aristotle realized it 2300 years ago that nature abhors a vacuum. Vacuums are being created as energy particles go by. A vacuum is just a relative state. Take a bottle of champagne. Vineyards that brew the stuff know that they must provide relatively stronger bottles for it; much stronger, for instance, than those used for the average bottle of red or white wine. The reason is that the process of making champagne can achieve a pressure of over 200 kPa within the bottle, double that of the ambient pressure outside of the bottle (101 kPa). When you pop the cork, you can see the huge burst of energy that is waiting there to immediately escape and equalize into the atmosphere. Without a strong bottle, the inside pressure would cause the bottle to explode.

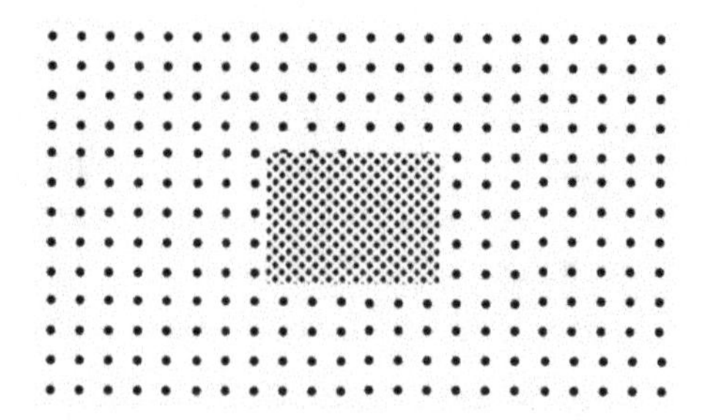

Figure 139 Vacuum: two adjoining spaces with different densities or pressures.

The phenomenon and effects of a vacuum are widespread. Here is an experiment that anyone can do, but be careful. If you stand next to a train track when a train is going by, even with your eyes closed, you will be able to feel the direction in which it is moving. When the last car goes by, you will notice a brief reversal of that flow. You can feel it. The train was going from left to right, but at the instant it has passed, the flow reverses briefly from right to left. That is because the energy of the train is creating a vacuum behind it and that vacuum will pull in particles until the two pressures equalize.[148]

Same with electricity flowing down a copper wire, it is flowing from left to right until the charge is dissipated and the flow is broken; for an instant thereafter, the vacuum it has created will cause a reversal of flow.

As for lightning, as the lightning charge goes down the lightning channel toward Earth, it creates a vacuum behind it, and energy tries to fill that vacuum. In this case, the lightning charge is actually ready to fall back into the vacuum it has just created. The lightning charge has strewn a vacuum behind it in space, and that energy is perfectly capable of turning around and coming back up the lightning channel. The huge energies of lightning create large vacuums, and energy has a terrific tendency to fill these vacuums causing the luminous event known as the return stroke.

Lightning's Downward Explosive Discharge (DED)

Cloud-to-ground lightning is an event where charge is transferred from the thundercloud down to the ground—hence its name. As shown in the previous section, the commonly accepted description of the return stroke mechanism is a misconception that can be held even by academics.

Google "return stroke" and you will be presented with things like this quote from a famous professor (who shall remain nameless) at an even more famous university (which shall also remain nameless): "*When the stepped-leader nears the ground the fields become so*

[148]Warning: the vacuum created by a speeding train can be strong. Many deaths are caused by trains pulling people onto the tracks after they have gone by. If you try this experiment, be sure to position yourself behind a strong railing or tie yourself securely to a tree or post. You can observe the same phenomena if you stand alongside a highway where container trucks are traveling by at speed.

strong that frequently a discharge will begin nearby at some elevated conducting object and jump up to meet the stepped-leader. Once a conducting path has been established between the ground and the cloud base, a surge of current called the return stroke moves up the channel defined by the stepped-leader..."

But what about the explosion?

Advancing a lightning model that fails to stipulate the downward explosive discharge (DED) as lightning's key element is like teaching the operation of a sailboat without mentioning the sail.

In 1915, German engineer and inventor, Dr. Wilhelm Schmidt, observed that lightning created actual explosions. In studying the "explosion waves" so produced, his equipment allowed him to estimate that although the greater part of the total lightning energy is transformed into heat and light, it is the long inaudible pressure waves in the immediate vicinity of the electrical discharge that cause the most mechanical damage.[149]

At the cusp between stepped-leader and return stroke is lightning's downward explosive discharge (DED). It should be labeled as such, describing as it does the "main event" of a cloud-to-ground lightning stroke. Of significance is the fact that the DED is the completion of a major cycle of action of the lightning strike. As we are dealing with electrical potential differences in the hundreds of megavolts, and as Earth ground is not a perfect conductor, an explosion of greater or lesser magnitude accompanies the neutralization of lightning charge.

As is true of any explosion, the force and energy of the DED rush out spherically in every direction. You can observe this in high-speed photos of every sort of explosion, from atmospheric nuclear explosions to conventional bomb blasts on the Earth's surface, to experiments performed in the science lab. See Fig. 140.

In the lightning explosion photographed by Tom Splietker (upper left box), only the top third of the sphere is visible. The percentage of lightning's force, energy, heat, radiation, etc. traversing back up the return stroke channel, though significant, would account for a very small part of the charge of the DED. Bottom right frame shows relative diameter of a return stroke.

[149]W. Schmidt, Thunder: theories and experiments conducted in an endeaver to solve the problem, *Scientific American*, **2045**, 1915, p.175.

Figure 140 Explosions propagate outward spherically. Upper left shows the top third of a lightning explosion. Bottom right shows the relative diameter of the ionized channel (about 1 to 20,000).

Although the DED may not have previously been fully appreciated or named, by no means was its function ignored. Prof. Uman refers to it several times: "When contact is made (by the upward streamer) with the stepped-leader, a violent, high-current discharge travels to ground."[150] And "When the stepped-leader contacts an upward-moving discharge (and through it is attached to the ground), negative charges at the bottom of the leader violently move to ground, causing large currents to flow and the channel of the stepped-leader near the ground to become extremely luminous."[151] Richard Feynman also commented on this phenomenon: "The moment the leader touches the ground, we have a conducting 'wire' that runs all the way up to the cloud and is full of negative charge. Now, at last, the negative charge of the cloud can simply escape and run out. The electrons at the bottom of the leader are the first ones to realize this; they dump out... So, all the negative charge runs out of whole column in a rapid and energetic way."[152]

The DED is different from the return stroke, and when we validate it as such, we open the door to a host of new discoveries and understandings about lightning.

---

[150]M. A. Uman, *Lightning*, Dover Publications, New York, 1984.

[151]M. A. Uman, Everything you wanted to know about lightning but were afraid to ask, *Saturday Evening Post*, May 13, 1972, p.37.

[152]F. R. Feynman, "Electricity in the Atmosphere" in The Feynman Lectures on Physics: mainly electromagnetism and matter—Lecture 9, Addison-Wesley Publishing Co., Reading, Massachusetts, 1964.

Polarities of Lightning

Benjamin Franklin was the first to correctly suggest the dual positive/negative nature of electrical charge. He was also the first to recognize lightning to be a form of that electricity. Over time, his theories have been validated by many experiments and much observation. And they have led to a host of further discoveries.

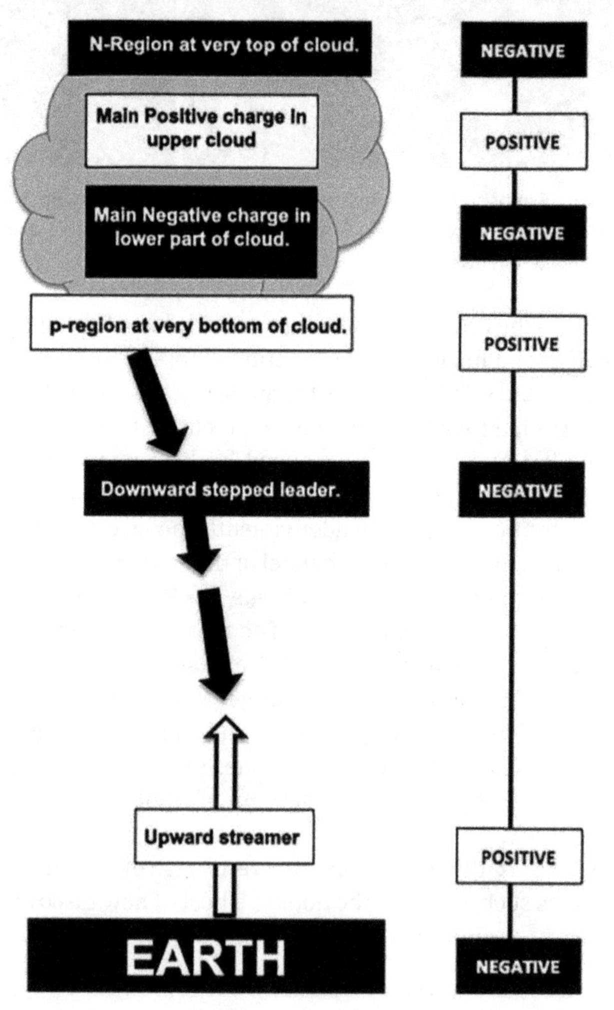

Figure 141 The main polarity reversals in most cloud-to-ground lightning.

As lightning became more carefully studied using instrumentation of increasingly probative power, more and more poles have been discovered to be intrinsic elements of the lightning process. E. R. Williams first showed it to be "tri-polar."[153] Figure 141 shows the main electrical field changes now commonly agreed to exist within the storm cloud itself.

The negative charge of the stepped-leader would account for pole 5 except for the fact that positive poles have been found to exist at the junction of each of the individual steps of a stepped-leader.[154,155] Assuming a stepped-leader with only five steps (in actuality, they can have many more), and adding the upward streamer and the negative charge of the Earth itself, the total number of pole changes in a single stroke of one negative cloud-to-ground lightning event comes to a minimum of 15. See Table 4. This pattern is no coincidence. These alternating poles are the woof and warp of the lightning process.

MIT Prof. Earle R. Williams joins Dwyer and Uman in pointing out that many aspects of lightning behavior have defied theoretical prediction and replication by models. While studying contemporary problems in lightning physics, Williams came to the realization that most of the outstanding ones are linked to differences in behavior of positive and negative polarity. This is obvious in such things as the ways negative and positive flashes differ in transferring their charge to ground. The negative flash does so by discrete multiple strokes, while the positive (usually) has only a single stroke followed by continuing current.

Williams makes the case that these differences are routed in what he calls "microscopic asymmetry in mobility for free electrons and positive ions."[156] He illustrates this idea with a diagram (Fig. 142) showing electrons converging at one end of an electric field and diverging on the opposite side.

---

[153]E. R. Williams, The electrification of thunderstorms, *Scientific American*, 1988, p.97.
[154]W. R. Gamerota, M. A. Uman, J. D. Hill, T. Nigin, J. Pikey, and D. M. Jordan, Electric field derivative waveforms from dart-stepped-leader steps in triggered lightning, *Journal of Geophysical Research Atmospheres*, **119**, 18, 2014.
[155]J. Howard, M. A. Uman, C. Biagi, D. Hill, V. A. Rakov, and M. D. Jordan, Measured close lightning leader-step electric field-derivative waveforms, *Journal of Geophysical Research*, **116**, D08201, 2011.
[156]E. R. Williams, Problems in lightning physics: the role of polarity asymmetry, *Plasma Sources Science and Technology*, **15**, 2006, S91–S108.

Table 4 Lightning's polarity shifts

Pole Change	Location	Charge
1	N-Region at extreme top of cloud	**Negative**
2	Main positive charge in upper cloud	**Positive**
3	Main negative charge of cloud	**Negative**
4	P-Region at very bottom of cloud	**Positive**
5	Downward stepped-leader: Step 1	**Negative**
6	Tip of stepped-leader: Step 1	**Positive**
7	Downward stepped-leader: Step 2	**Negative**
8	Tip of stepped-leader: Step 2	**Positive**
9	Downward stepped-leader: Step 3	**Negative**
10	Tip of stepped-leader: Step 3	**Positive**
11	Downward stepped-leader: Step 4	**Negative**
12	Tip of stepped-leader: Step 4	**Positive**
13	Downward stepped-leader: Step 5	**Negative**
14	Upward streamer	**Positive**
15	Earth	**Negative**

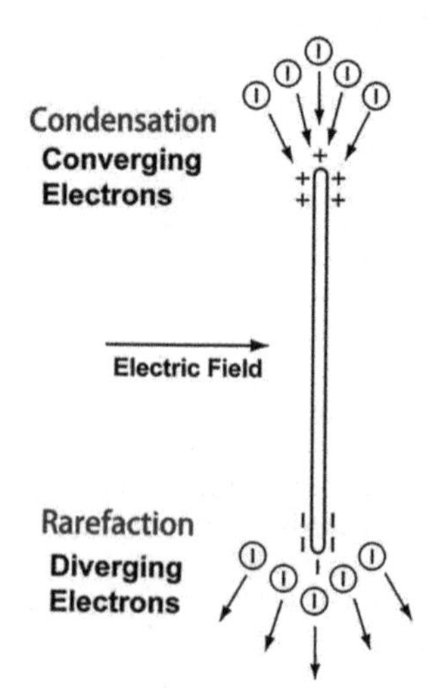

Figure 142 Electrons in a condensation/rarefaction wave (Williams).

Figure 143 is a more concrete example showing an airplane being hit by lightning—a somewhat unusual occurrence but cited here because it shows the physical mechanism of polarity change so well.

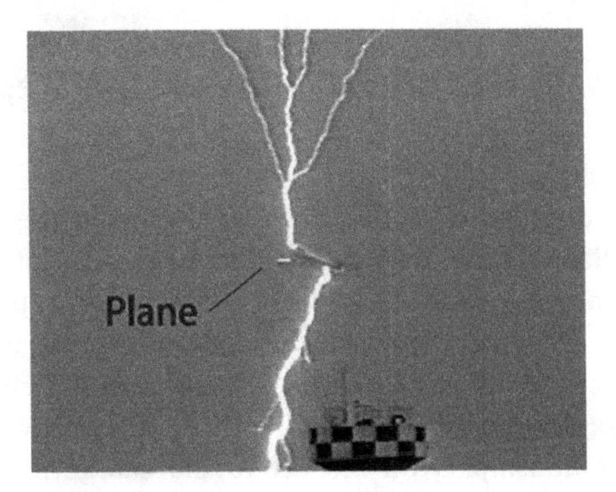

Figure 143 Airplane hit by lightning.

You can see that the lightning propagates upward above the point of the plane and downward below it. Although not possible to determine the specific polarities from the photo alone, it is a certainty that polarity asymmetry is a decisive factor in this photograph. A useful question that could be asked is what function did the plane play in producing that change in polarity?

What might be as important as the behavior differences of the positive and negative poles is the underlying mechanism creating those poles. The identification of the physical process lying behind these alternating shifts in polarity would go a long way toward solving the mysteries of lightning.[157]

The Stepped-Leader

There could be no DED without the creation of an ionized channel between the cloud and the ground on which the charge could flow. The stepped-leader is the mechanism that creates that channel in a series of rapid luminous steps about 50 yards (or 50 m) long—hence

[157]See data in next section on ridges.

its name. Step by step it builds the channel that will eventually span the distance between cloud and ground. At that precise moment when the connection is made, the DED transfers the energy from cloud to ground. Polarity change is one of the main factors.

Problems with Existing Models of the Stepped-Leader

The charge distribution of the downward moving stepped-leader cannot be directly measured, and models have so far fallen short in explaining the stepped-leader process. Some models of the stepped-leader have assumed a uniform distribution of charge along the leader channel.[158] Others postulate a linear change in charge density from the thundercloud to the leader tip.[159] Yet others have suggested complex relationships with respect to the charge dissipated by the return stroke.[160] D. Djalel et al., in studying the currents and electromagnetic fields computed by these various model types, pointed out that none of them could be reality checked since current measurements are only possible at the channel base, and "none of the models take into account the attachment process so they probably do not very accurately model the electromagnetic field at the early parts of the lightning discharge."[161]

After studying all available models, Mason concluded that none of the theories put forward in the past 100 years "was capable of generating and separating charge at a rate sufficient to account for the observed frequency of flashes and the changes of electric field accompanying them."[162]

---

[158]R. H. Golde, *Lightning*, Academic Press, London, 1977.

[159]V. Cooray, V. Rakov, and N. Theethayi, The lightning strike distance: Revisited, *Journal of Electrostatics*, **65**, 296, 2007.

[160]B. F. J. Schonland, The pilot streamer in lightning and the long spark, *Proceedings of the Royal Soceity of London A*, **220**, 25, 1953.

[161]D. Djalel, H. Ali, and C. Fayçal, The return-stroke of lightning current, source of electromagnetic fields, *American Journal of Applied Sciences*, **4**, 1, 42–48, 2000.

[162]J. Mason and N. Mason, The physics of a thunderstorm, *European Journal of Physics*, **24**, 2003, p.S101.

Malan and Schonland Tips

Schonland and Malan give us a straightforward and uncomplicated look at the process of the stepped-leader. Schonland provided the diagram in Fig. 144 showing the stepped advance of an initial negative leader. The dark area between A and B is the previous (completed) leader-step. D. J. Malan discerned three facts about stepped-leaders that are as enlightening today as they were back in 1963.[163]

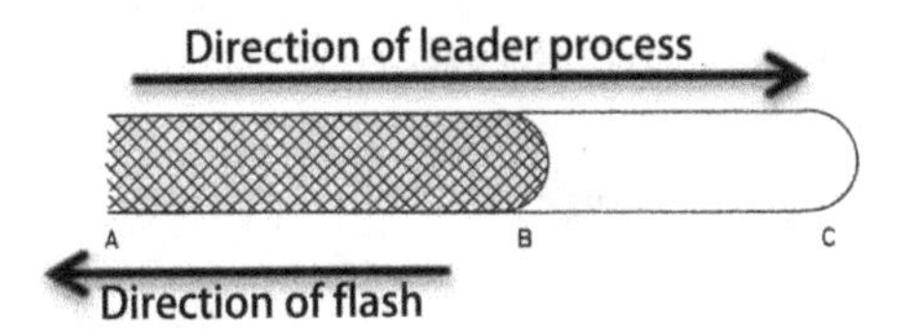

Figure 144 Schonland diagram illustrating stepped advance of the initial leader. A-B is last completed step. B-C is the next step.

1. There is a short but distinct "quiescent" period between each of the steps of the stepped-leader (Point B in Fig. 144). At that point, the electrical field drops below some threshold and progress stops "until something happens."
2. The tip of each negative step (B) develops a positive charge.
3. At the instant of initiation of the new negative step, the glow condition changes to an arc condition and an explosion occurs that illuminates the point at the tip of the step (B) as well as a considerable length of the channel behind it. (From B back to A and beyond...)

What is happening at the intersection of each of those steps? It is as though each step of the stepped-leader has its own personal "return stroke." Perhaps it can be partially glimpsed in the photo of the plane. Could the charge hitting the barrier be creating a reversal of polarity? The next sections address the exact mechanism.

Lightning is Electromagnetic Radiation

That lightning is or generates electromagnetic radiation is not a new discovery. Thirty years ago, Williams described it as follows: "Once

[163]D. J. Malan, *Physics of Lightning*, English Universities Press, London, 1963.

a thundercloud has become charged to the point where the electric field exceeds the local dielectric strength of the atmosphere—that is, the strength of the atmosphere to support a separation of electric charge—a lightning flash results... During that fraction of a second the electrostatic energy of accumulated charge is transformed into electromagnetic energy."[164]

Lightning emits waves that vibrate at almost every part of the electromagnetic spectrum from the megahertz of radio waves through the gigahertz of microwaves, including the terahertz of visible light all the way up to the exahertz of X-rays and gamma rays. This places lightning firmly in the category of an electromagnetic energy source. This is important when considering the propagation of lightning charge.

Propagation of Electromagnetic Radiation: Classical View

In the 180 years, up to and including the first quarter of the 20th century, various theories to explain the propagation of light and electromagnetic energy were introduced. What they all had in common was the foundation of an electrical wave moving through a medium. The wave theory of light (by which light propagated in spherical waves through a medium) was a discovery of Christiaan Huygens, Dutch mathematician and scientist, who published his "Treatise on Light" in 1678.[165] Two hundred years later, the father of lightning science, Benjamin Franklin, fully endorsed this theory.

Textbooks from Franklin's time all the way up through the first quarter of the 20th century explained, in clear and straightforward terms, that energy was propagated in waves through a medium with little or no differentiation made between light waves and sound waves. See footnotes 36 and 37. Basic concepts of lightning became muddied around 1938 as a result of several confusions, the main ones being: (1) Was lightning a particle or a wave, and (2) did it need a medium through which to propagate?

---

[164]E. R. Williams, The electrification of thunderstorms, *Scientific American*, 1988, p.97.
[165]C. Huygens, *Treatise on Light*, The Hague, written in 1678, rendered into English by S. P. Thompson, University of Chicago Press, 1912.

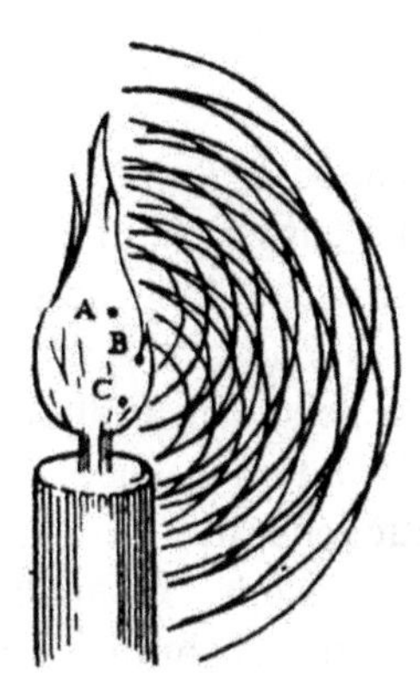

Figure 145 Huygens propagation of light through a medium.

Lightning Propagation

Dr. Wilhelm Schmidt, in 1915, explained the charge transfer process of lightning in terms of shock waves, condensations, and rarefactions.[166] According to Schmidt, waves propagate through the particles of an ionized channel by creating a series of condensations and rarefactions. When the wave compresses the particles, a density or condensation is achieved (the dark areas in Fig. 120). Areas where the particles are thin are the "rarefactions." Figure 142 (Williams) shows this characteristic—charge moving horizontally in a series of steps as electrons are condensed by the waves in the conducting channel. This mechanism accords very well with the stepped-leader process.

A. W. Smith was equally clear that lightning was an electrical wave moving through a medium. "*In all cases of wave motions it is necessary to have a medium in which the waves can travel...Wave motion is characterized by the handing on from place to place in an elastic medium of disturbances emanating from some source. These disturbances follow each other at definite intervals and the wave motion thus maintained transfers energy from one point to another. The individual particles in the medium move to and fro about their normal positions of rest and the waveform only moves forward. A medium is necessary for the transmission of the disturbance and a*

[166]W. Schmidt, Thunder: theories and experiments conducted in an endeaver to solve the problem, *Scientific American*, 2045, 1915, p.175.

definite time is required for the disturbance to travel from one place to another."[167]

Rakov and Uman chose to begin their seminal book, *Lightning: Physics and Effects*, with a basic definition of the lightning process from D. J. Malan, which is at once descriptive and functional: "An electrically active thundercloud may be regarded as an electrostatic generator suspended in an atmosphere of low electrical conductivity."[168] That definition reduces lightning to two basic elements: an "electrostatic generator" (a device capable of producing a high voltage) and a transmission medium of relatively high impedance. To simplify further, lightning is an electrical discharge flowing through a medium.

The electrons in a lightning channel do not move very far. Uman notes that "individual charges are not lowered over the relatively large distance from cloud to ground during the relatively short time duration of the lightning discharge. Rather...any flow of electrons (the primary charge carriers) into...the top of the lightning channel results in the flow of other electrons in other parts of the channel, much as would be the case were the channel a conducting wire. Thus, coulombs of positive or negative charge can be effectively transferred to ground during the time that an individual electron in the channel moves only a few meters."[169] CIGRE confirmed this in a 2013 report.[170] As for the electrons, Uman reiterates, "None go very far. All charge transfer is 'effective'."[171] "Effective" is amplified by Prof. Rakov: "You may think of effective charge transfer as a kind of domino effect."[172]

Almost every significant advance ever made in the subject of electricity was made against the specific backdrop that electromagnetic waves propagated through a medium. But by the 1930s, textbooks started to say that electromagnetic waves (including

[167]A.W. Smith, *The Elements of Applied Physics*, McGraw-Hill Book Company, New York, 1923, p.250.

[168]D. J. Malan, Physics of the thunderstorm electric circuit, *Journal of the Franklin Institute*, **283**, 6, 1967.

[169]M. A. Uman, *The Lightning Discharge*, Academic Press, Orlando, 1987, p.10.

[170]CIGRE Technical Bulletin 549, *Lightning Parameters for Engineering Applications*, Paris, August 2013.

[171]M. Uman, personal communication to author.

[172]V. Rakov, personal communication to author.

lightning) radiated and propagated "without the intervention of a material substance."[173]

Today our students are taught that "electromagnetic waves require no material medium to exist,"[174] but just because that opinion appears in thick textbooks does not make it so. Lightning charge propagates in waves, down ionized channels in air, with the free electrons being the conductors.

Virgin Air

To coalesce with quantum mechanical theories, modern papers on lightning have invented a new term: virgin air. Not only is this not a scientific term, it is not even correct English. Yet it has become a foundation of much of the current lightning research. For example, in a 2016 peer-reviewed survey of 33 papers on the subject of lightning propagation, the term "virgin air" was used 22 times without ever defining it.[175]

This is a subject not often discussed in connection with lightning, but it should be because just about every luminary in the history of electricity has rejected any such concept. Jean-Jacques Ampère, Benjamin Franklin, Heinrich Rudolf Herz (first person to prove the existence of electromagnetic waves), Lord Kelvin, Hendrik Lorentz, James Clerk Maxwell, and Nikola Tesla all subscribed to one or another concept of a medium through which electromagnetic waves propagated. Unfortunately, as shown earlier, and to the detriment of our students and researchers in lightning science, this concept is no longer taught in universities.

The Condensation/Rarefaction Process

Understanding lightning's condensation/rarefaction process requires an understanding of energy, a subject addressed in the

---

[173]R. J. Stephenson, *Exploring in Physics*, University of Chicago Press, Chicago, Illinois, 1935, p.96.

[174]J. Walker, *Fundamentals of Physics* (Halliday & Resnick, 10th Ed.), Wiley & Sons, Hoboken, N.J., 2014, p.445.

[175]M. D. Tran and V. A. Rakov, Initiation and propagation of cloud-to-ground lightning observed with a high-speed video camera, *Scientific Reports*, **6**, 39521, 2016, doi:10.1038/srep39521.

earlier section of this book. We shall not repeat that material here, except to say that energy is simply **the movement of particles or impulses between points in space** and this definition, held by every great inventor since Sir Isaac Newton, was challenged in the early 1900s, and finally abandoned in the 1930s.

Using the preceding definition, energy can be seen to manifest three major characteristics: flows, dispersals, and ridges. See Figs. 112 and 146. In these characteristics, we can find all known energy phenomena, including the propagation of lightning, gamma rays, X-rays, visible light, plus many other phenomena such as boundary layers, hydrophilic states, exclusion zones, galvanic action, and surface tension.

Dispersals are explosions emanating out spherically 360 degrees from a single point. Ridges can be understood as condensations of matter or energy brought about by the actions of flows and dispersals hitting up against each other and creating an enduring state of matter.

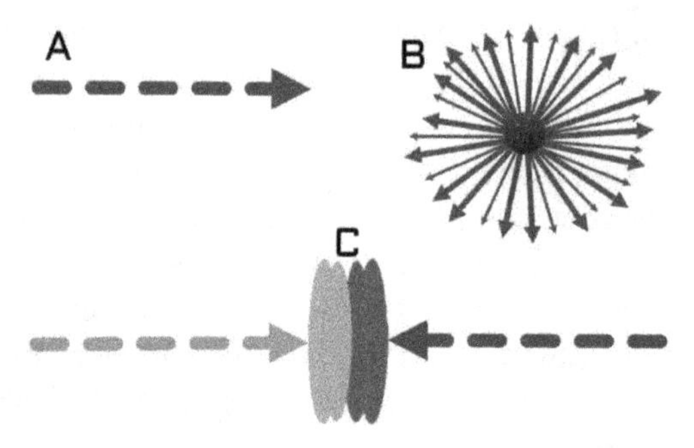

Figure 146 The three basic energy characteristics: (A) a flow, (B) dispersal (explosion), and (C) opposing energy flows creating a ridge.

The same is true for lightning in air. A path must first be created before energy can propagate down it. That path may be only a few centimeter in diameter.[176] It is the creation of that path—a bounded ionized space between the potential charge differences in the cloud and the ground—that is the basic function of the stepped-leader. Figure 147 shows the progress of a series of condensations and rarefactions and their consequent polarity changes at each step of the stepped-leader.

[176]M. A. Uman, *Lightning*, Dover Publications, New York, 1984.

With the understanding that energy is transferred through a medium, we can proceed to look at the process by which a lightning charge traverses between cloud and ground. We do this by resurrecting the classical interpretation of this process: condensation/rarefaction. The lightning wave propagates through the particles of a medium by a series of condensations and rarefactions. When the wave compresses the particles, you achieve density or a "condensation." When that density reaches a certain threshold (within a few microseconds), a small explosion (dispersal) occurs. As some areas get denser, the areas that have lost particles become rarefied (fewer particles per unit volume of space). These are the rarefactions. See Figs. 147 and 148.

Figure 147 Stepped-leader process. A series of condensations and rarefactions flow in steps from cloud to ground. Right side of image is a magnification of the polarity shifts.

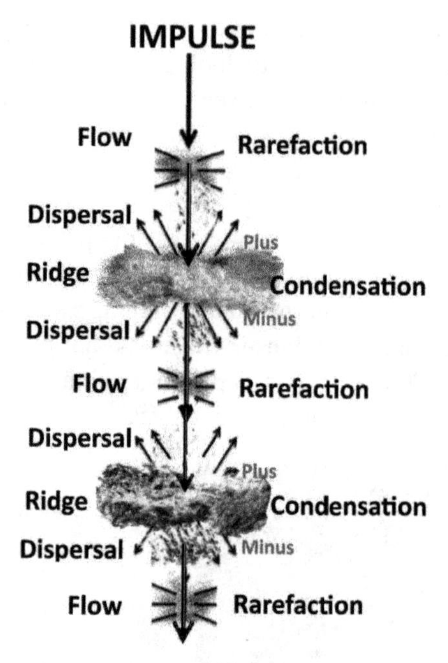

Figure 148 What the particles in a lightning strike are doing.

The frequency of lightning is such that the cycle of flow → dispersal → ridge can be repeated many times a second. This mechanism also accounts for the polarity changes in lightning.

Conclusion

It is a worthy goal to get to the bottom of lightning's better-kept secrets, and it is suggested to those who wish to succeed in this endeavor that they reassess some of the more recent theories of lightning such as "return stroke" and "propagation in virgin air" and base their inquiries on more fundamental principles of physics such as those discussed in this book.

The explanations of lightning held by science before 1938 were straightforward and transparent. When scientists were allowed to hold these views, great understandings were being achieved in lightning science. Utilizing them once again is a surefire way of improving our penetration into those mysteries of the lightning flash that have so far eluded understanding.

Mathematics Postscript: Proofs for the Energy Miracle Keys[177]

Since no one seems to object very loudly when quantum mechanics declare real things imaginary and imaginary things real, I shall take some of that creativity onto myself and declare this section not part of this book. (It could not be part of the book, since it includes mathematical formulas and I have promised to include no mathematics in the book proper.)

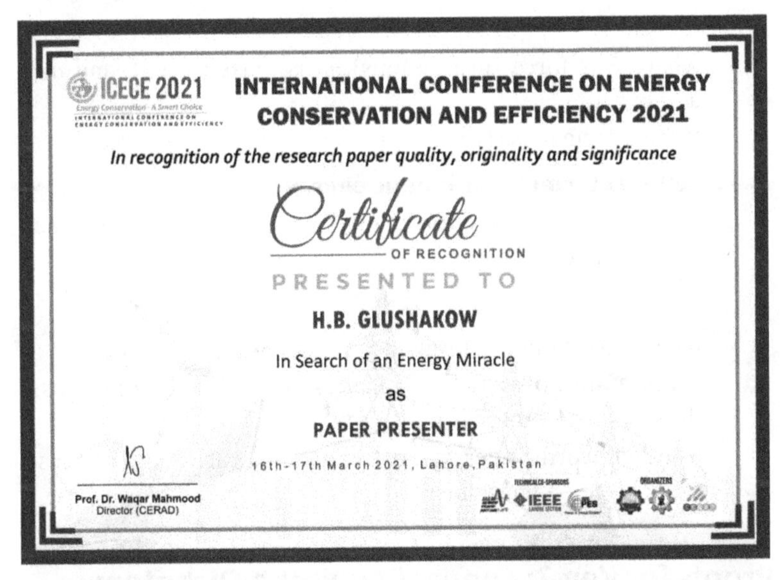

Figure 149 Certificate awarded by the 2021 International Conference on Energy Conservation and Efficiency (ICECE) in recognition of the originality and significance of the material contained in this section.

What follows are some of the most important and valuable mathematics in the history of energy and electricity—mathematics on which the greatest breakthroughs in energy were based, and which prove the correctness of the keys to the Energy Miracles listed in Chapter 6.

[177]Most of the data in this section were included in a peer-reviewed article entitled "In Search of an Energy Miracle," presented at the International Conference on Energy Conservation and Efficiency (ICECE), an IEEE-sponsored event held in March 2021.

Proofs for Key 1: Energy Consists of Postulated Particles in Space

Energy consists of the motion of particles in space. The similarity of the force and energy formulas proves this. Anyone who has experienced a lightning strike close hand will attest to the incredible amount of force it contains.

1. **Newton's second law confirms this**

   Derived from Sir Isaac Newton's work, this formula is most commonly expressed as

   $$F = ma$$

   where F is force (in Newtons), m is mass (in kg), and a is acceleration.

   If there is no mass (i.e., m = 0), there would be no force.

2. **Leibnitz formula for kinetic energy**

   The Leibnitz formula for kinetic energy (KE) also expresses this concept:

   $$E_K = \frac{1}{2} mV^2$$

   where E_K is kinetic energy, m is mass, and V is velocity.

   If there are no particles or if those particles have no mass, then $m = 0$ and there is no energy.

   In other words, energy must consist of particles and these particles must have mass.

Proofs for Key 2: Energy Requires a Dichotomy

1. **Maxwell's equations**

   All energy production comes about from dichotomies. Maxwell's equations prove this for both electricity and magnetism.

   His first equation (Gauss' law) can be written as:

   $$\nabla \cdot E = \frac{\rho}{\varepsilon_0}$$

where ∇ is divergence, E is electric field, and $\frac{\rho}{\varepsilon_0}$ is the charge density. The plus (source) and minus (sink) of an electric field are the bases of that field. Without either the plus or the minus, $\nabla = 0$ and there is no charge.

Similarly, his second equation is

$$\nabla \cdot B = 0$$

Gauss's law for magnetism, where B is the magnetic field, shows that if both a positive pole and a negative pole do not exist, then $\nabla = 0$. This proves that no magnetic monopole can create energy. Both plus and minus poles are needed.

2. **Faraday's Law**

   This is one of the basic laws of electromagnetism and explains why electric generators work. It tells you that changes in an electric field will create changes in a magnetic field (and vice versa):

$$\nabla \times E = -\frac{\partial B}{\partial T}$$

   where ∇ is divergence, E is the electric field, and $-\frac{\partial B}{\partial T}$ is the change in magnetic flux density. The plus and minus of the electric field produce the electric field (E). It requires both a plus and a minus. If either is missing, then there is no electric field, and hence no change in magnetic flux.

Proof for Key 3: Energy Generation Requires a Base

Coulomb's law

In any electrical generator or energy-producing device, a base must be established to keep the two poles/terminals of the dichotomy apart. This is easily seen in a battery where the base keeps the positive and negative terminals separated, but it is also true for any system or device where you want to encourage or create the motion of particles (energy) and is proven by Coulomb's law:

$$F_e = \frac{kq_1q_2}{r^2}$$

where F_e is the electric force, kq_1q_2 is the charge, and r is the distance between two charged objects.

Distance must be maintained between at least two objects for electrical energy to be produced. If r equals zero (as is the case of a single object where the distance between it and itself is zero), no energy is created.

Proof for Key 4: Energy Requires a Medium through Which to Propagate

Ohms law

$$V = IR \text{ or } R = \frac{V}{I}$$

where V is voltage (difference in potential between two objects), I is electrical current, and R is resistance (the relative ease or difficulty an electrical current has in moving through a medium). R is always a function of an actual medium. For example, the resistance of a solid copper tube 1 m long is lower than the resistance of the same size tube made of rubber. (The resistance of rubber is hundreds of thousands of times greater than that of copper and can be measured.) Ohm's law works in all cases except where $R = 0$. If resistance is zero, there can be no difference in potential between two objects and thus no current.

In other words, where there is no medium, there can be no energy potential or energy flow.

This ends the Mathematics Postscript.

Epilogue

Is There More to the Universe Than the Mechanics of Space, Energy, and Time?

What do you think? There most certainly is. There's YOU. It was humorous to find a few quantum mechanics paralyzed with fear over the prospect that they might be able to create effects at a distance. This ability has been proven many times by many people. It has either happened to you, or you surely know a person who has thought about someone he had not seen for a long time only to have that person up and phone him "out of the blue" or send him a message. Walk by a person 50 feet away or even drive by in a car and put your attention on him. If you are really good, he will look up at you. Many a person has become immediately aware of the death or serious threat to a loved one even though it occurred thousands of miles away.

There are mechanics, and there is life. By mechanics, we mean the subjects covered in this book: the energies, matter, motions, spaces, and time. Of primary importance to any mechanical scheme is space. Then energy. Next is the solidified or condensed energy we call matter. And finally, always present in any mechanical arrangement is that relative change in position of particles in space known as time. The elements of mechanics are all quantitative, meaning there is so much distance, so much mass, or so many hours. Mechanics do not create and do not destroy. The entire subject of physics is based on the conservation of mass. Throw a table into a fire and watch it burn up. Where did it go? Not far. If you collect up all the gases, energies, and ash by-products of the burning, you will have all the parts of the original table. They have just moved around a bit and altered their form.

Not so with life. You are life. Life has quality and ability. And life can create. The ability of life is demonstrated by its handling of the mechanics of space, energy, and time. Life can consider and have opinions and make decisions for the future. Life is aware of its own awareness. Space, energy, time, and form are the by-products of life and are monitored and controlled by life. Life is an entirely different subject, and being outside the realm of physics, it has no place in this book.

In this book, we have been looking at mechanical laws, not the laws of life. The rules of life may be different but do not cancel Newton's laws. They do not change anything. Finding out about them can make life more interesting, more fun, and more rewarding. Have you lived before this life? Are poltergeists really real? Maybe only those who have run into one "in the flesh" will vouchsafe for them. But real or not, they do not change the laws of science, so you do not have to worry about them.

What is life, really? What are you really? The world's only scientific definition of life was advanced by Ron Hubbard who divined that life is basically a static. As such, it has no mass, no motion, and no wavelength. But the amazing thing is that even without a location in space or in time, it has the ability to postulate and perceive.[178]

Climate Epilogue: Is It Already Too Late?

While researching for this book, I visited the largest bookstore in one of America's busiest shopping malls to check out its Climate Change section. There wasn't one. Disappointing, but not a deal breaker. Of course, it is not too late. Human self-generated resource can yet save the day if humans who care and want to make a difference start looking in the right direction.

The discovery of Energy Miracles and the resulting elimination of fossil fuels will produce miraculous changes. Planetary warming and rising seas would be reduced and existing resources of air and water cleaned up. For the first time, it would be within the reach of

[178]See Axiom 1. *Advanced Procedures and Axioms* by L. Ron Hubbard.

humans that everywhere their people could have access to plentiful sources of clean energy, water, and food.

There can be another not so obvious, but equally important result. For at least 5000 years, humans have never had to come together as "humans" to save the world, their people, and environments from such a universally grave threat.

A military unit is only a group of mistrustful individuals until it has been under enemy fire and made it out alive. What was just a band of men before becomes a band of brothers, a team whose members can now rely on each other and trust each other. That is what it takes to make a **great** military unit.

It will serve the planet well for its peoples and countries to come together at this time in common cause. The consequences would extend far beyond 2050 and the discovery of a few Energy Miracles, but the successful quest for the Energy Miracles could be the catalyst that sets this process in motion.

I have never met Mother Nature, but if I ever do, I will tell her:

> *"Look, Ma'am. It doesn't matter how many storm surges you throw at our continents, how many glaciers you melt, or how many feet of your seawater you send to flood our islands and coastal cities. It also doesn't matter what is your political affiliation, or whether you're just exhibiting a misguided show of feminism. It's irrelevant whether global warming is being entirely caused by you or whether we ourselves have been inadvertently helping speed the planet closer to catastrophe. None of that matters because we will not be sitting idly by and letting mankind perish. If we can put men on the moon, send ships to Mars and Saturn, and harness the atom, we can certainly preserve our beachhead on planet Earth."*

The Author

H. B. Glushakow, Senior Member of the Institute of Electrical and Electronic Engineers (IEEE), has been active in the power quality and alternative energy fields for 25 years. During this time, he has written several dozen books, peer-reviewed technical papers, and White Papers plus lectured in 20 countries on six continents. He holds a patent for a system that protects mega wind turbines from the damaging effects of lightning. A trenchant website he created prompted the European-based International Electrotechnical Commission (IEC) to adopt a Code of Ethics for the first time in its 114-year history. He lives with his wife, Sharon, part of the time in their apartment overlooking the Chaoyang Park in Beijing, China, and the rest of the time in their home on the Withlacoochee River in the U.S.

Figure 150 The author on a visit to the Tesla Technical Museum in Zagreb, Croatia.

Review Quotes

H. B. Glushakow tackles the most important issue of our time with wit and intellect. He illuminates the history of energy and the sudden stall in its progress for the past 100 years. The narrative is as deep as it is poignant, a union of facts and science. The author walks through the fundamentals of energy technology and sets the stage for the only solution left on the issue of global warming, an energy miracle. The stage is methodically set with the turning of each page. In the end, the reader feels the urgency, understands the risk, but is also inspired by a sense of hope and optimism that fuels this author at his very core. Well timed, well-crafted book. One that should interest all those interested in the future of humanity on this planet.

—**Lee Rudow**, President and CEO of Transcat, USA

By identifying the roadblocks in the way of inventing a new source of clean and affordable energy, this book makes a great contribution to solving humanity's biggest problem, global warming. Many people do not dare look into the eyes of global warming's inconvenient truth, but this author clearly shows where science has gone astray and in which directions the solutions lay. His writing style is down to earth—efficient for scientists as well as easy for the average person to read. The illustrations are funny and informative and very welcome. If his ambition was to 'cut through the noise' and give readers the clarity needed to tackle this subject, he has succeeded. I see this book as an invitation to everyone, to face the facts, understand the problems, and participate in the common goal of solving this crisis. Of the many books I have read on climate change this one gives us the best chance of solving it, and while reading it, I felt the urge to become a part of shaping the future along the guidelines proposed by the author.

—**Matej Simonič**, President of TESTING, Slovenia

I enjoyed the narrative, content as well as the perspective. I wish you success and widespread sales of this book.

—**François D. Martzloff**, National Institute of Standards and Technology, USA

People who know me have been shocked hearing the 180-degree turnaround on climate change I've undergone after reading an advanced copy of Energy Miracles: The Global Warming Backup Plan. *I dare say what I learned reading this book has turned me from "an infuriating, closed-minded climate change skeptic" to "a passionate climate-change advocate." If you know anyone who is as stubbornly block-headed as I was just 8 days ago, urge them to order a copy of this masterpiece while the planet still has a chance to avert its current tragic trajectory. If they won't buy it for themselves, buy it for them. That's just how important I now know this crisis truly is, and thanks to this book, just how fixable it still is, given the ingenious technical and political solutions the author prescribes to avert the impending disaster before it's too late for us.*

—**W. Gould**, wealth manager

This book presents an impassioned and cogent case for the need to push forward (by looking back) in order to discover the energy miracle that Earth (not to mention mankind) needs for its very survival.

—**C. Place**, general reader

Thoroughly enjoyed reading it. The book's observation that we don't teach electrical anymore is true. We teach how to steer the electrons around once we have electricity, but there is no emphasis on new ways to create those moving electrons. Great stuff!

—**Martin Woodruff**, Science Instructor, UK

Index

Printed in the USA
CPSIA information can be obtained
at www.ICGtesting.com
LVHW010020080823
754481LV00005B/236